T0369103

Synthesis Lectures on Data Management

This series publishes lectures on data management. Topics include query languages, database system architectures, transaction management, data warehousing, XML and databases, data stream systems, wide scale data distribution, multimedia data management, data mining, and related subjects.

Eduardo Ogasawara · Rebecca Salles ·
Fabio Porto · Esther Pacitti

Event Detection in Time Series

Eduardo Ogasawara
CEFET/RJ
Rio de Janeiro, Brazil

Fabio Porto
LNCC
Petrópolis, Rio de Janeiro, Brazil

Rebecca Salles
INRIA, CNRS, LIRMM
Université de Montpellier
Montpellier, France

Esther Pacitti
LIRMM and INRIA
University of Montpellier
Montpellier, France

ISSN 2153-5418 ISSN 2153-5426 (electronic)
Synthesis Lectures on Data Management
ISBN 978-3-031-75940-6 ISBN 978-3-031-75941-3 (eBook)
https://doi.org/10.1007/978-3-031-75941-3

This Springer imprint is published by the registered company Springer Nature Switzerland AG
The registered company address is: Gewerbestrasse 11, 6330 Cham, Switzerland

To our parents:
Ana Maria and Tsuneharu
Neudes Meire and Adolpho
Edmê and João André
Eunice and Tércio

and to our families:
Maria Izabel and Maria Eduarda
Neudes, Adolpho and Samella
Patrícia, Tiago, and Daniel
Patrick and Anna

Preface

Time series analysis is a cornerstone across various domains, including finance, healthcare, environmental science, and cybersecurity. As time series increase in velocity, variety, and volume, so does the challenge of extracting meaningful events within them. While events can be classified into anomalies, change points, and motifs, event detection is the process of identifying these events. It is a usual practice to study anomalies, change points, and motifs separately. Some books address these research areas isolated, but the literature lacks a broad view of event detection.

This book, *Event Detection in Time Series*, fills this gap by providing fundamental and state-of-the-art methods for detecting events in time series. It bridges theory and practice, presenting key concepts alongside practical examples with accessible code for real-world problems. Whether you are a researcher, practitioner, or student, this book guides you through the core concepts, methods, and challenges in event detection.

The structure of this book reflects the multifaceted nature of the subject. We begin with an introduction covering the definition of these three different types of events. Then, we explore time series analysis methods and data preprocessing techniques that form the foundation for event detection. Subsequent chapters delve into the fundamentals of anomalies, change points, and motifs. Later on, the following two chapters are driven by online event detection and evaluation metrics. The last chapter summarizes the overarching challenges and future directions. We also provide an appendix covering Harbinger, our framework to support time series event detection, so readers can quickly start practicing. We hope this book will inspire further research and development in this subject.

We are indebted to the many researchers whose work has laid the groundwork for the advances discussed in this book. We also acknowledge the support of our institutions, colleagues, and families, who have made this work possible. First, we thank Prof. Patrick Valduriez for his deep and outstanding review. He gave us many advice, which enriched our writing a lot. We also thank our partners from Inria, mainly Florent Masseglia and Reza Akbarinia, who helped us to develop maturity on the subject. We also thank Profs.

Dayse Pastore, Diego Carvalho, Eduardo Bezerra, Joel Santos, and Rafaelli Coutinho for your time reviewing the book.

We also thank many students and former students who support us in early using or implementing either materials or methods available at Harbinger, including Ana Beatriz Cruz, Antonio Castro Filho, Antonio Mello, Cristiane Gea, Diego Salles, Ellen Paixão, Heraldo Borges, Lais Baroni, Luciana Escobar, Janio Lima, Jessica Souza, Lucas Tavares, Leonardo de Carvalho, and Murillo Dutra.

Finally, the authors thank CNPq, CAPES, and FAPERJ for partially sponsoring this research.

Rio de Janeiro, Brazil — Eduardo Ogasawara
Montpellier, France — Rebecca Salles
Petrópolis, Brazil — Fabio Porto
Montpellier, France — Esther Pacitti

Contents

Introduction 1

1.1 Overview

Time series analysis focuses on examining the entirety of the data collected over time. However, in many practical scenarios, it is essential to highlight and examine specific sequences of observations, known as events [29]. Time series events are instants or intervals in the time series where observations change in a manner that is considered important for analysis or decision-making processes. The interpretation of an event can vary significantly across different domains. For example, in weather time series, events could be intervals of heavy rainfall or severe storms, while in economic time series, significant market shifts, such as sudden drops or gains, are important events. Events can also mean a security breach, unusual network traffic spikes, a machine breakdown, or a system failure that disrupts normal operations in an industrial setting [214].

In information theory, time series events can be understood as intervals marked by significant entropy changes, suggesting that these segments contain more information than others. This definition allows analysts to identify and quantify the impact of these events more precisely [31]. Furthermore, time series events can be classified into three main types, each with distinct characteristics: anomalies, change points, and motifs. Anomalies are observations that differ significantly from typical observations, change points indicate significant shifts in the time series trend or volatility, and motifs are recurring approximate sequences in the time series. These types are explained and formalized in Sect. 1.2.

Each type of event requires specific analytical methods and tools for detecting and interpreting events. Time series event detection systematically identifies intervals representing significant deviations from typical observations or predefined conditions, i.e., detecting intervals with a significant change in the entropy. The goal is to pinpoint the exact moments or periods of their occurrences [304]. Detecting events in time series requires studying

E. Ogasawara et al., *Event Detection in Time Series*, Synthesis Lectures on Data Management, https://doi.org/10.1007/978-3-031-75941-3_1

advanced and specialized methods. This topic has been extensively explored over the years and has increasingly attracted more interest over the last decade [39].

To get started defining these concepts, let us provide some basic formalization. A time series X is a sequence of n observations, $<x_1, x_2, x_3, \ldots, x_n>$, where x_1 and x_n represent, respectively, the observations of the series at the first (oldest) and last (newest) instants [95]. The length n of a time series X is $|X|$. A specific time series observation is referenced as x_t, indexed in time by t ($t \in \{1, \ldots, n\}$). It is common to interpret the sequences that compose a time series as a time series process [259].

A time series is studied as a function of past observations [129]. When observations are related to a single variable, a time series is referenced as univariate. Conversely, it is a multivariate time series when observations are related to two or more variables.

The frequency of a time series defines when the observations are regularly collected. For instance, a yearly time series gives the observations collected each year. This case characterizes a low-frequency time series when observations are collected daily or slowly. Conversely, high-frequency time series are those in which the collection of observations occurs more than once during the day.

Consider the Yearly Global Temperature Time Series (YGT) depicted in Fig. 1.1. It is an example of a univariate time series. It is available from the National Centers For Environment Information (NOAA), which provides comprehensive data collections of global coverage over land (Global Historical Climatology Network-Monthly) and ocean (Extended Reconstructed Sea Surface Temperature) surface [206]. YGT has been available since 1850 and is relevant to the discussion about global warming [204]. YGT is useful for presenting the outcomes of event detection methods in the following sections.

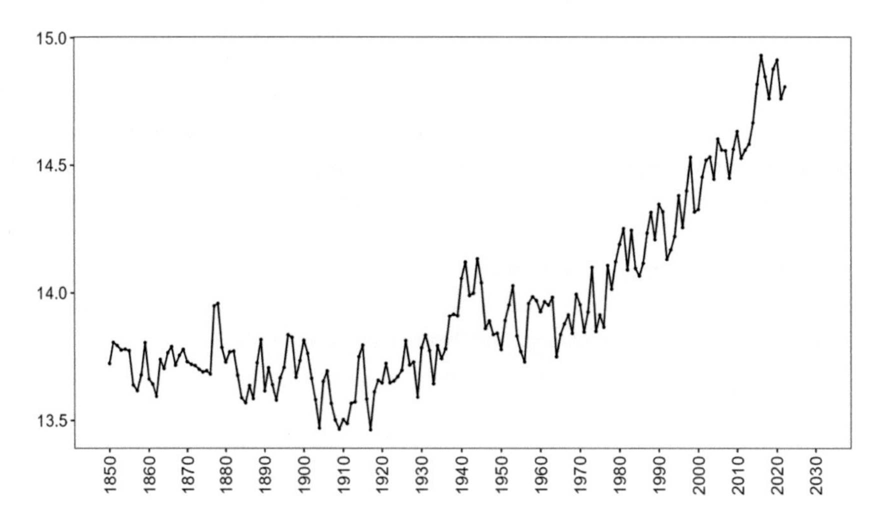

Fig. 1.1 YGT (in degrees Celsius) obtained from NOAA [206]

Event detection can be organized according to how the time series are processed, i.e., identifying past events (offline), discovering events in real time (online), and even predicting future events. It is a basic function in surveillance and monitoring systems, gaining much attention in application domains involving critical systems [228]. The event detection problem is targeted in applications based on sensor data analysis [76], which can be observed in chemistry, reflection seismic, and oil drilling and exploration, where monitoring operations are essential. Furthermore, event detection is important for practical applications that affect day-to-day life, such as medical diagnosis, disease outbreaks, fault detection, structural damage identification, intrusion detection, fraud detection, wireless networks, and astronomy object detection.

The event detection methods in the literature adopt five general groups: regression, classification, clustering, statistical, and domain-based. Unfortunately, assessing event detection methods is complex [325]. Thus, many authors have analyzed, compared, and reviewed several methods [212].

Event detection in time series is a wide and dynamic area of research. However, most papers focus on particular aspects of the problem, such as detecting only a specific type of event [281]. Furthermore, they do not generally formalize events and their different types. Although online event detection is becoming increasingly important [191, 207], most event detection algorithms do not address streaming applications [325].

This chapter introduces, using a taxonomy that we have developed, the principles associated with event detection, which are analyzed according to event type specializations, event detection methods, and evaluation metrics. We formalize point events as a generalization for anomalies and change points and cover the definition of motifs and discords. The taxonomy provides a context for the reviewed literature (see Fig. 1.2) and was built based on a systematic search of relevant papers. First, papers on event detection for time series were searched using a query string involving the keywords "event detection" and "time series". A second search was performed to find papers on the specific area of event prediction using the keywords "event prediction" and "time series". Furthermore, survey papers on event detection and event prediction were gathered based on snowballing.[1] Finally, an additional snowballing search was performed on detection performance metrics due to their increasing importance.

The taxonomy is divided into five major categories. The first two categories present the type of events and the general data structure associated with them. The following categories group the events according to their detection scenarios, methods, and performance evaluation of detection methods. These categories are detailed in the next sections.

[1] Snowballing, also known as citation chaining, is the process of searching the references (backward) and the citations (forward) of a paper to identify other relevant papers.

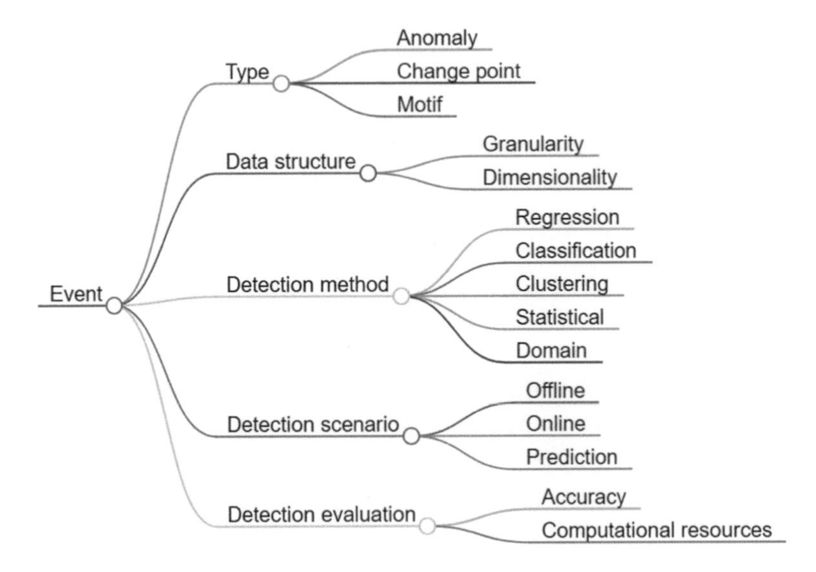

Fig. 1.2 Taxonomy of time series events

1.2 Types of Events

Events are pervasive in real-world time series. As mentioned before, they can be characterized as anomalies [212], change points [281], motifs (i.e., repeated sequences in the time series) [252], and discords (i.e., atypical sequences in the time series) [311]. The first two commonly occur in an instant. Events related to intervals are generally studied as motifs or discords [177]. Event detection methods are mostly specialized in identifying a specific type of event. When some methods can detect multiple events in time series [27], they are related to detecting anomalies and change points [63].

Let us start with a general formalization of point events. Let x_t be observations for a time series X. A Temporal Component (TC) for x_t is expressed as $tc(x_t)$. The TC can refer to the observation itself, its instant trend, and its instant volatility, respectively, represented as x_t, $tr(x_t)$, and $v(x_t)$. Consider a typical time series in which the autoregressive assumption holds [119]. A TC for x_t is expected to relate relatively to previous observations. Such assumption can be expressed as $ep(tc(x_t), k) = E(tc(x_t) \mid tc(x_{t-1}), \ldots, tc(x_{t-k}))$, where ep is the expected (E) TC for x_t from the previous k TC. Using the same assumption, it is expected that a TC for x_t can be explained from the following observations. It is similar to the autoregressive assumption applied to a reversed time series [173]. Thus, a given $tc(x_t)$ is related to the following k observations. Such assumption can be expressed such as $ef(tc(x_t), k) = E(tc(x_t) \mid tc(x_{t+1}), \ldots, tc(x_{t+k}))$, where ef is the expected (E) TC for x_t from the following k TC.

Point events $\{t\}$ in a time series X can be expressed as $e(X, k, \sigma)$ using Eq. 1.1, where k represents the length of nearby observations and σ is a tolerance threshold. Suppose an TC for x_t escapes the expected value above σ based on the previous or following k observations. In that case, it can be considered an event. Equation 1.1 also considers an event if the expected TC from the previous and following k observations differ above a threshold σ.

$$\begin{aligned} e(X, k, \sigma) = \{t, & |tc(x_t) - ep(tc(x_t), k)| > \sigma \\ & \vee |tc(x_t) - ef(tc(x_t), k)| > \sigma \\ & \vee |ep(tc(x_t), k) - ef(tc(x_t), k)| > \sigma\} \end{aligned} \tag{1.1}$$

1.2.1 Anomalies

Anomalies are observations that do not conform to the typical ones in the time series [243]. Anomalies and outliers are often interchanged [121], but some authors make a slight difference between them, relating outliers to the statistical perspective of the data distribution [289]. An outlier can be defined as an observation (or subset of observations) that appears inconsistent with the remainder of the observations. It is compatible with the chosen data distribution, albeit very uncommon [212]. On the other hand, anomalies seem to obey a different distribution than the typical observations [212]. Since this distinction is blurred, for the sake of simplicity, they can be understood as similar concepts with the caveat that anomalies are events that we intend to detect, while outliers are usually unwanted observations that we intend to discard.

Anomalies can be caused by various factors, such as errors in data collection, changes in underlying trends or patterns, or external factors, such as weather events or economic shifts [326]. They appear as unexpected spikes, dips, or irregular patterns indicating critical incidents, such as system failures, economic events, or emergent behaviors in complex environments [219]. Anomalies can be formalized as in Eq. 1.2. In this case, an event identified in x_t can be considered an anomaly if it escapes expected TC before and after instant t according to the σ threshold.

$$a(X, k, \sigma) = \{t, |tc(x_t) - ep(tc(x_t), k)| > \sigma \wedge |tc(x_t) - ef(tc(x_t), k)| > \sigma\} \tag{1.2}$$

Noise Anomalies

Noise anomalies, or residual anomalies, are the simplest type of anomalies. They correspond to unexpected or abnormal variations in the residual component of a time series, which is the part of the time series that is not considered by the seasonal and trend components (see Chap. 2 for time series components and time series decomposition).

Noise anomalies can be formalized as follows. Let $\hat{X}$ be an estimate of a time series X produced by adjusting a model α, with $\hat{x}_t = \alpha(x_t)$. From $\hat{X}$, it is possible to derive the residual time series W ($< \omega_1, \ldots, \omega_n >$), with $\omega_t = x_t - \hat{x}_t$. When the model is well-adjusted, it is expected that W has a Gaussian distribution with zero mean (in this case, it is considered a Gaussian white noise) [283]. Noise anomalies (na) can be identified using statistical distribution tests (sdt) over the residual time series W, as described in Eq. 1.3.

$$na(X) = sdt(W), \ \omega_t = x_t - \hat{x}_t \tag{1.3}$$

The simplest sdt assumes that W follows a Gaussian distribution and is a parametric analysis. Considering that $\overline{W}$ and $sd(W)$ are, respectively, the mean and standard deviation of W, anomalies are observations that are below $\overline{W} - 3 \cdot sd(W)$ or above $\overline{W} + 3 \cdot sd(W)$ [125]. Equation 1.4 characterizes the instants in which observations do not conform to the Gaussian distribution.

$$sdt(W) = \{t, \omega_t \notin [\overline{W} - 3 \cdot sd(W), \overline{W} + 3 \cdot sd(W)\} \tag{1.4}$$

A more general solution for sdt does not assume that W follows a particular distribution (nonparametric analysis). In this case, box plot analysis is adopted and can be described using Eq. 1.5, where $Q_1(W)$ and $Q_3(W)$ are the first and third quartiles, respectively, and IQR is the interquartile distance [6]. The equation characterizes the instants where observations are atypical in a box plot analysis.

$$sdt(W) = \{t, \omega_t \notin [Q_1(W) - 1.5 \cdot IQR(W), Q_3(W) + 1.5 \cdot IQR(W)]\} \tag{1.5}$$

Figure 1.3 shows the occurrence of noise anomalies (marked as red) from the YGT, considering the residual time series obtained from an adjusted ARIMA model (ARIMA is covered in Chap. 2). As can be observed, such anomalies correspond to local minimum

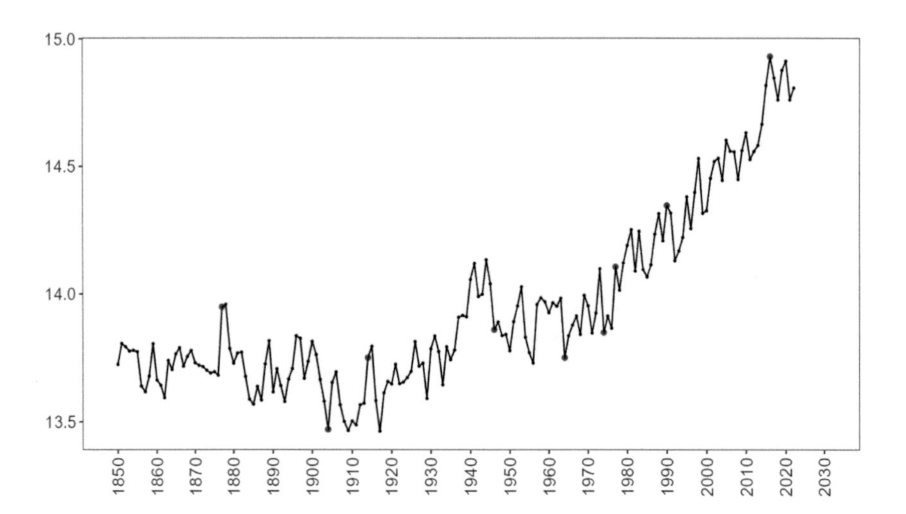

Fig. 1.3 Anomalies observed at YGT from the residual analysis of an ARIMA model

and maximum values concerning the modeled time series. Anomalies are fully addressed in Chap. 3.

1.2.2 Change Points

Change points are time intervals in a time series where there is a significant change in the statistical properties, e.g., changes in mean, variance, correlation, or other parameters that characterize the time series distribution [112]. They represent a transition between different states in a process that generates the time series [272]. In this case, a change point event identified at time t follows the expected TC observed before or after t, but not both at the same time according to $cp(X, k, \sigma)$ in Eq. 1.6. It can also refer to a significant difference between the expected TC before and after instant t.

$$\begin{aligned} cp(X, k, \sigma) = \ & \{t, (\ |tc(x_t) - ep(tc(x_t), k)| > \sigma \veebar \\ & |tc(x_t) - ef(tc(x_t), k)| > \sigma\ ) \\ & \vee\ |ep(tc(x_t), k) - ef(tc(x_t), k)| > \sigma\} \end{aligned} \tag{1.6}$$

Change points are also related to concept drift [294], in particular, when considering multivariate time series D, such that X ($X \subset D$) is a predictor to Y ($Y \in D$). It is possible to define change point detection as a concept drift hypothesis test. The null hypothesis H_0 characterizes the absence of drift, and the alternative hypothesis H_A negates H_0. Formally, H_0 and H_A are as in Eq. 1.7, where $\mathbb{P}_{n(t)}$ is the probability density function for D of observations nearby $n(t)$ [23]. Such observations could be a sliding window (see Chap. 2 for definition) containing t.

$$\begin{aligned} H_0 &: \forall\, t, k\ (t \neq k) \mid \mathbb{P}_{n(t)} \approx \mathbb{P}_{n(k)} \\ H_A &: \forall\, t\ \exists\, k\ (t \neq k) \mid \mathbb{P}_{n(t)} \neq \mathbb{P}_{n(k)} \end{aligned} \tag{1.7}$$

From this perspective, change point detection is related to monitoring concept drifts. Figure 1.4 shows the occurrence of change points (marked as dashed gray lines) from the YGT considering using the Chow test statistic [319]. As observed, trends before and after change points are significantly different. Change point detection is deeply addressed in Chap. 4.

1.2.3 Motifs and Discords

Some events in time series can also be observed by the presence of *motifs* and *discords*. Time series motifs are sequences of significantly similar observations within a time series. They correspond to recurring patterns with some distinctive shape in the time series [177]. In other words, a motif is an approximately repeated subsequence within a longer time

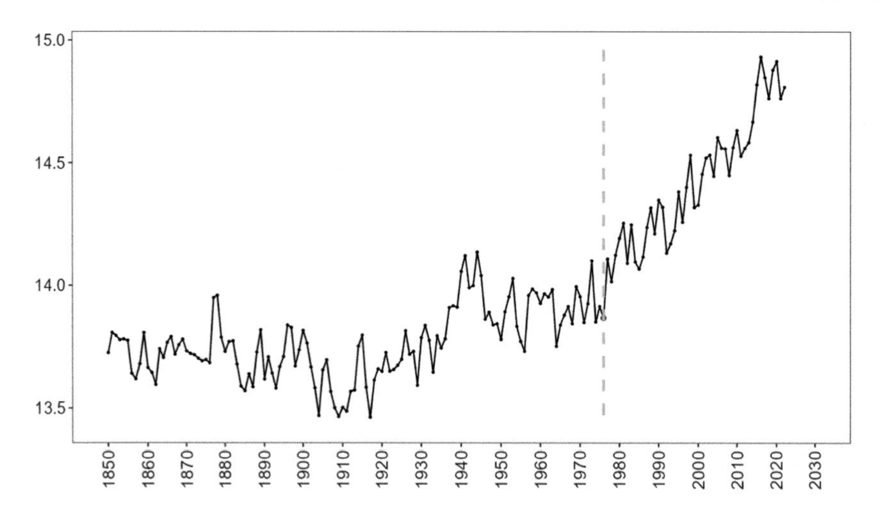

Fig. 1.4 Change points observed at YGT using Chow test statistic [319]

series. Conversely, *discords* are sequences that do not repeat in the time series. They are significantly different sequences from the remainder of the observations.

Motifs can be used to understand the underlying structure of a time series and identify important features such as trends, seasonal patterns, or outliers [279]. They become more interesting whenever these sequences provide more information than typical time series observations. In these cases, they do not occur so many times. Identifying motifs can be useful in various applications, including anomaly detection, classification, and predicting future observations.

Figure 1.5 shows a small fraction of a patient's heartbeat in an electrocardiogram. It is available at MIT-BIH [227] and contains many similar sequences due to regular heartbeats. These common subsequences are motifs. However, three irregular heartbeats are depicted in Fig. 1.5. They correspond to labeled sequences annotated by specialists related to arrhythmia signals. The beginning of each sequence is marked in green, and the following observations of each sequence are marked in purple. The interesting question is how to discover them automatically. Furthermore, they can be a motif if interpreted as three similar rare subsequences or as three discords if considered significantly different.

Given a sequence q and a time series X, q is a *motif* in X, if and only if q occurs in X at least σ times [44]. The general method for discovering motifs is related to directly exploring time series distances or correlations between subsequences. Such brute force exploration yields quadratic complexity $O(n^2)$. Thus, indexing-based methods based on Symbolic Aggregate approXimation (SAX) [176] and hashing have been used to improve computation time while introducing similarity tolerance for the sequences. Recently, state-of-the-art methods, such as matrix profile [312], have adopted enhanced data structures for fast computation. The formalization of motifs and discords is presented in Chap. 5.

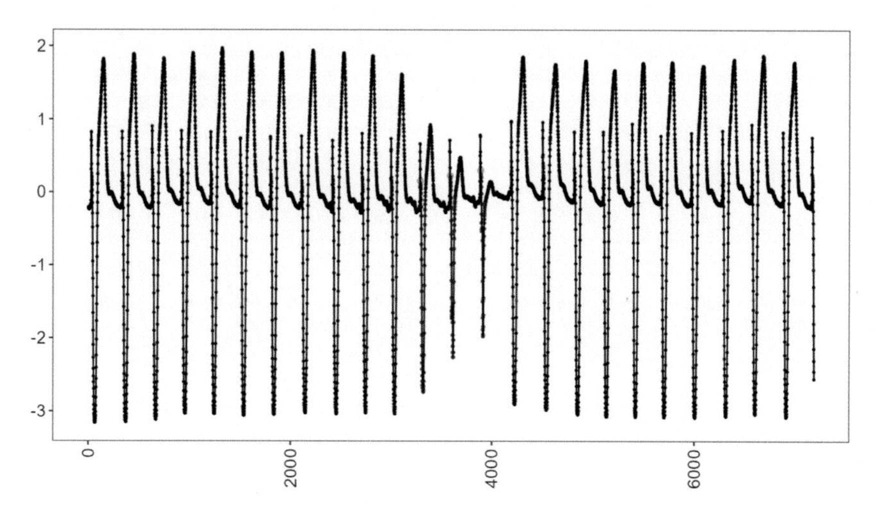

Fig. 1.5 Three labeled sequences of an electrocardiogram. They might correspond to a motif with three occurrences or three different discords

1.3 Data Structure

Both time series and events can be represented in various data structures depending on the context and purpose of the analysis [146]. One common data structure for time series is a simple table or matrix, where each row represents a specific time, and each column represents a variable or feature being measured at that time. Under this structure, an additional column can represent the presence of events, with the characterization of the type of event. It is a typical format for many time series and associated events [146].

Another common data structure for time series is a specialized object or class in a programming language or statistical software package. These objects are specifically designed to handle time series and often have built-in functions and methods for manipulating and analyzing the data. Examples include the Pandas library in Python or the TS object in R. In addition to these basic structures, more advanced techniques such as decomposition or state space models can represent and analyze time series, particularly when dealing with multiple variables or complex dependencies over time [259].

1.3.1 Granularity

Events fall into two categories: point and interval [32]. They are related to the granularity of the time series. Point events are individual observations interpreted as events concerning the rest of the time series. In other words, they are discrete events recorded at a particular timestamp or time interval. Point events are represented in a timestamped event data structure.

year\month	1	2	3	4	5	6	7	8	9	10	11	12
1971	13.9	13.8	13.8	13.8	13.9	13.8	13.9	13.9	13.9	13.8	13.9	13.9
1972	13.7	13.7	14.0	14.0	13.9	14.0	14.0	14.0	13.9	14.0	14.0	14.0
1973	14.3	14.3	14.3	14.2	14.1	14.2	14.1	14.0	14.0	14.0	13.9	14.0
1974	13.8	13.7	13.9	13.9	13.9	13.8	13.9	14.0	13.9	13.8	13.8	13.8
1975	14.0	14.0	14.0	14.0	14.0	14.0	13.9	13.9	13.9	13.8	13.7	13.8
1976	13.9	13.8	13.8	13.9	13.8	13.9	13.9	13.9	13.9	13.7	13.9	14.0
1977	14.1	14.1	14.2	14.2	14.2	14.2	14.1	14.1	14.1	14.0	14.1	14.0
1978	14.1	14.1	14.1	14.0	14.0	14.0	14.0	13.9	14.0	14.0	14.1	14.0
1979	14.0	13.8	14.1	14.0	14.1	14.1	14.1	14.2	14.2	14.2	14.2	14.4
1980	14.3	14.3	14.2	14.2	14.3	14.2	14.2	14.1	14.1	14.1	14.2	14.1

(a)

year	value
1971	13.8
1972	13.9
1973	14.1
1974	13.8
1975	13.9
1976	13.9
1977	14.1
1978	14.0
1979	14.1
1980	14.2

(b)

Fig. 1.6 Examples of global temperature time series under different granularity between 1971–1980: **a** Monthly global temperature (MGT); **b** Yearly global temperature (YGT)

This structure stores information about each event, including the time it occurred and any associated data.

Interval events refer to events that occur over a defined period rather than at a specific time. They are recorded as a start and end time along with the time series. Examples of interval time series events include continuous daily rainfall or high hourly electricity consumption.

Analyzing interval events requires different methods than those for analyzing point events. One approach is to aggregate the data into larger time intervals and then analyze these aggregates using point event methods, such as the previously mentioned ARIMA. The YGT presented so far is derived from the Monthly Global Temperature Time Series (MGT). Figure 1.6 shows a sample of the MGT. Figure 1.6a shows the MGT from January 1971 until December 1980. Figure 1.6b shows the YGT for the same period. It is temporal aggregated from the MGT. The value for the YGT is the mean of MGT each year.

1.3.2 Dimensionality

Event detection might occur in univariate or multivariate time series. A univariate time series is a time series in which only one variable is measured in each observation. This variable can be a continuous or discrete numeric value measured at regular or irregular intervals. Univariate time series events are represented as a sequence of values over time. The simplest data structure used to represent it is a vector. Although simple, univariate time series events are important as they can provide valuable insights into various phenomena.

A multivariate time series is one in which multiple variables are measured in each observation. Each variable can be continuous, discrete, or categorical values. Examples of multivariate time series events include a city's hourly temperature, humidity, and wind speed readings. Multivariate time series events are represented as a matrix or table with multiple columns, each column representing a measured variable and each row representing an obser-

value
13.8
13.9
14.1
13.8
13.9
13.9
14.1
14.0
14.1
14.2

(a)

time	value
1971	13.8
1972	13.9
1973	14.1
1974	13.8
1975	13.9
1976	13.9
1977	14.1
1978	14.0
1979	14.1
1980	14.2

(b)

time	global temperature	crude oil production
1971	13.8	2491
1972	13.9	2634
1973	14.1	2870
1974	13.8	2875
1975	13.9	2740
1976	13.9	2966
1977	14.1	3069
1978	14.0	3108
1979	14.1	3229
1980	14.2	3111

(c)

Fig. 1.7 Examples of time series dimension representation: **a** univariate time series as vector; **b** univariate time series with timestamp column; **c** multivariate time series

vation or measurement taken at a specific time. Unlike relational tables, the order of data is relevant. Figure 1.7 illustrates the representation of different time series, where Fig. 1.7a shows the univariate YGT as a vector. Figure 1.7b shows the YGT with an additional column to characterize the time associated with each observation. Figure 1.7c shows a multivariate time series, with a column for the YGT and the worldwide crude oil production. One might be interested in finding associations among the different dimensions in a multivariate time series, e.g., if an association exists between crude oil consumption and the global temperature increase.

1.4 Detection Methods

Several methods for event detection, including Machine Learning (ML) and big data processing, have been described and compared [32, 60]. In addition to the five general groups (regression, classification, clustering, statistical, and domain-based), event detection methods can be classified into theory-driven or data-driven. Theory-driven models are based on established theories, principles, or concepts from a particular field. These models aim to represent and explain real-world phenomena by explicitly incorporating theoretical constructs into the model. Conversely, data-driven models are created by analyzing large time series. While theory-driven methods are mostly in domain-based and statistics-based analysis, ML generally enables data-driven methods.

Data-driven methods based on ML are not necessarily restricted to certain kinds of problems and do apply to time series event detection. Some of the main data-driven methods used for event detection are K-Nearest Neighbors Conformal Anomaly Detector (KNN-CAD), K-Means, Neural Network (NNET), Support Vector Machine (SVM), Extreme Learning Machines Network (ELM), Convolutional Neural Network For Time Series (Conv1D), and

Long Short-Term Memory Neural Network (LSTM). In particular, NNET, ELM, SVM, Conv1D, and LSTM are used for time series prediction [202].

Data-driven methods can also be organized according to how they learn from data. In supervised learning, the methods assume the availability of a training dataset with labeled instances (for example, the indication that each observation has an associated event or not). In semi-supervised learning, they assume that the training data only label instances for the typical observations. On the other hand, methods that operate in unsupervised learning do not require labels in training data and, thus, are most widely applicable. The methods in this category implicitly assume that typical instances are more frequent than anomalies [60].

Domain-based methods use expert knowledge or established models of the domain problem to identify new data samples that are not typical in case they differ enough from what is expected. Examples in this group include a specialized time–frequency method using domain-specific theory for finding events in a seismic time series [105], or the identification of human fall events based on a model of body posture evolution [321].

1.5 Detection Scenarios

The detection scenarios are an important characterization of the condition in which the event detection methods are used. They drive the requirements for data management, computational resources, and constraints of the event detection methods. We can organize the detection scenarios into three major classes: offline event detection, online event detection, and event prediction.

Offline event detection is when events are discovered after the time series has been collected. It involves analyzing the time series retrospectively to identify patterns or changes that may indicate the occurrence of an event. The process of offline event detection involves two major steps. First, the time series is preprocessed. Then, a time series analysis method is applied to identify events. Once potential events have been identified, further analysis is done to confirm whether they are real events or false positives. This analysis can involve additional data sources, expert knowledge, or further statistical testing.

Online event detection is when events are discovered in a time series as the observations are collected. It requires continuous monitoring of the time series. The process comprises the two major steps depicted in offline detection. However, it has computational constraints to enable real-time reactions when potential events are discovered. For example, predictive maintenance can detect changes in a machine that may indicate a potential failure, allowing maintenance to be scheduled proactively.

More formally, let X be a streaming time series composed of a continuous stream of inputs: $< \ldots, x_{t-2}, x_{t-1}, x_t, x_{t+1}, x_{t+2}, \ldots >$. At each time t, a model trained on previous observations $< \ldots, x_{t-2}, x_{t-1} >$ is used to determine whether the current observations of the system are unusual. This determination must be made by the next input x_{t+1}. For this, the time series model must be continuously updated [9].

Unlike offline event detection, data is not split into static train/test sets, and algorithms cannot look ahead. Practical applications impose additional challenges. When the sensor streams are large in number or when data comes at high velocity, there is little opportunity for expert intervention, manual parameter tweaking, and manual data labeling. Thus, operating in an unsupervised, automated fashion is necessary [9]. Furthermore, in streaming applications, early detection of events is often important, where events are identified as soon as possible. Detection of events can provide information that is important to decision-making. This information must be given early enough that it is actionable, preventing possible critical system failures. However, there is still a tradeoff between early detections and false positives. An algorithm that often makes inaccurate detections is likely to be ignored.

Finally, time series event prediction forecasts future events based on historical time series. It uses statistical, ML, or deep learning methods to model the patterns and trends in the time series and make predictions about future events [45]. The process involves several steps. First, the time series is preprocessed and cleaned to remove any noise or outliers that could affect the accuracy of the analysis. Then, various time series analysis methods are applied to model the patterns and trends in the data. These methods can be from ARIMA to deep learning models such as LSTM.

Once the model has been trained and validated, it can be used to predict future events. Time series event prediction is used in various applications, including finance and healthcare. In healthcare, for example, it can be used to forecast patient outcomes or the spread of diseases.

1.6 Detection Evaluation

The evaluation of event detection methods is related to determining the effectiveness of the algorithms, which differ according to the computational time and memory used.

1.6.1 Accuracy

The evaluation of time series event detection assesses the accuracy and effectiveness of event detection algorithms in identifying events in time series. It involves comparing the output of an algorithm with ground truth data, which experts have manually labeled to indicate the occurrence of events. The evaluation of event detection algorithms is challenging due to the subjective nature of event labeling and the lack of ground truth labels in many applications. The evaluation involves several metrics, which are calculated using the ground truth data and the output of the algorithm to evaluate the accuracy of the event detection algorithm. These metrics include precision, recall, F1-score, and area under the receiver operating characteristic curve (AUC-ROC). Precision is the ratio of true positive events to the total detected events. At the same time, recall is the ratio of true positive events to the total number of actual events. The F1-score is a weighted average of precision and recall that considers both metrics. The AUC-ROC measures the algorithm's ability to discriminate between positive and negative events [245].

An important aspect of any event detection method is how the events are reported. The outputs produced by event detection methods are either scores or labels. Scoring detection methods assign an anomaly score to each instance in the data, which varies according to the degree to which that instance is considered an anomaly. On the other hand, labeling detection methods assign a label (typical or anomalous) to each data instance [264].

Detection methods may have different performances under different time series [98], which calls for performance comparison. However, benchmarking and comparing event detection performance is hard [215]. Standard classification metrics are generally used to evaluate the performance of an algorithm to distinguish between typical and atypical data samples that can be directly adopted for point events. However, many real-world event occurrences extend over some time and have an interval. Therefore, some authors expand the well-known Precision and Recall metrics to assess the accuracy of detection algorithms for interval events [276].

1.6.2 Computational Resources

Event detection methods vary according to the computational resources required. The complexity of time series event detection can be analyzed using several metrics, including time complexity, space complexity, and computational complexity. Time complexity is the time required to execute the algorithm on a given input size. In contrast, space complexity refers to the amount of memory the algorithm requires. Computational complexity refers to the resource requirements, including time, memory, and other computational resources.

Several methods can be used to manage the complexity of time series event detection. They include optimizing the algorithm and its implementation, using parallel processing to distribute the computational workload, and using data reduction techniques to reduce the size of the time series.

1.7 Book Structure

The taxonomy in Fig. 1.2 lists the main concepts analyzed in depth in the following chapters. Figure 1.8 gives the general structure of the book. Chapter 2 provides the basics of time series analysis. It covers time series decomposition, time series preprocessing, and time series prediction, which are supporting methods for event detection.

Chapter 3 introduces the categories of anomalies and describes the main anomaly detection methods. Chapter 4 introduces the categories of change points and the main change point detection methods. It also covers concept drift in time series. Chapter 5 presents time series indexing and similarity, which are the main enabling concepts for motif discovery.

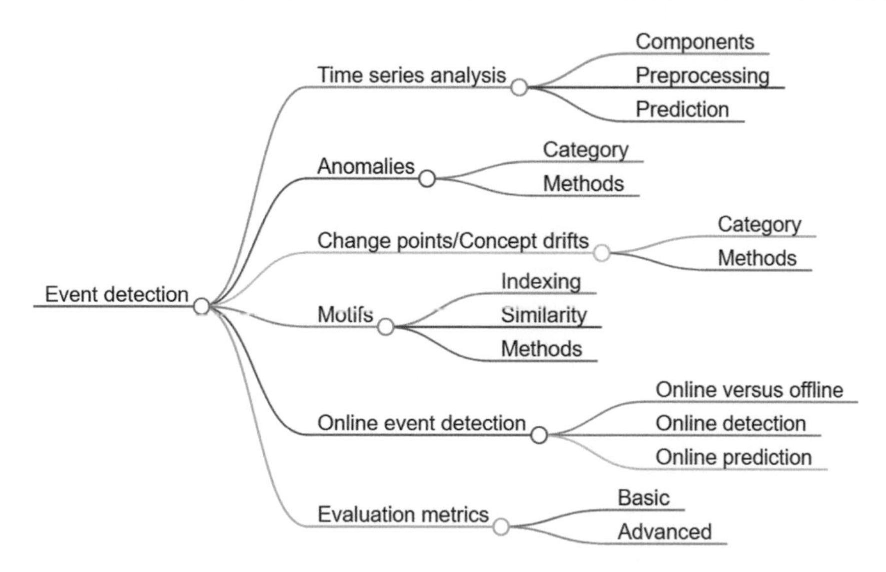

Fig. 1.8 Book structure

Later, the chapter covers the main motif discovery methods. Chapter 6 deals with online event detection. It starts by comparing online with offline detection. Then, it distinguishes between online detection (detecting events as soon as possible) and online prediction (predicting events before their occurrences). Chapter 7 covers the main evaluation metrics on event detection. It also presents time tolerance and specific event metrics, such as sequences and online event detection. Finally, Chap. 8 summarizes the research area and the main challenges presented in the book.

The book also includes Appendix A that describes Harbinger, which is our framework for event detection.[2] Harbinger is open source and available as an R package. All examples presented in this book were built on it and can be run using it.

[2] Available at https://cefet-rj-dal.github.io/harbinger.

Time Series Analysis

2

2.1 Time Series Components

Time series can be studied according to their major components, organized into three types: *(i)* trend (β_t); *(ii)* seasonality (π_t); *(iii)* residual (ω_t) as described by Eq. 2.1 [139].

$$x_t = \beta_t + \pi_t + \omega_t. \tag{2.1}$$

The trend component refers to the time series's long-term movement or direction. It represents the underlying pattern or trend of the data, independent of any seasonal or cyclical fluctuations. When the trend is linear, it is a monotonic function or a function in which there can be at most one extreme within a given data period [195]. The trend is taken as a disposition of the time series, which remains in future observations [100]. The trend can be rising, falling, or stable over time. For example, a country's gross product may steadily increase over several years, indicating a positive trend. A neonatal death rate may steadily decrease over time, indicating a negative trend. Both these situations are important for policymakers. Thus, the relevance of positive or negative trends is much more related to the semantics of the studied time series rather than its values [139].

Accurate analysis requires identifying and accounting for trends in time series. There are several methods to determine the trend component. The simplest is through linear regression, which involves choosing an average line for the time series. Other methods, including moving averages and more advanced ones, are presented later in this chapter. It is a common practice to remove the trend component of the time series (in a process called detrending) to enable focusing on the remaining time series components [119].

Seasonality refers to patterns that repeat at fixed intervals, such as daily, weekly, monthly, or yearly. These patterns usually relate to periodical factors influencing the underlying data. For example, sales may increase every December due to Christmas, or beverages and ice cream sales may increase during summer. Seasonality is also expressed according to

E. Ogasawara et al., *Event Detection in Time Series*, Synthesis Lectures on Data Management, https://doi.org/10.1007/978-3-031-75941-3_2

amplitude, frequency, and phase. Amplitude refers to the magnitude of seasonal variation. Frequency refers to the length of the seasonal cycle. Finally, the phase refers to the timing of seasonal peaks and troughs within each cycle [139].

The residual is the difference between the time series and the sum of the trend and seasonal components. The residual component represents the random or unpredictable fluctuations that the trend or seasonal components cannot explain. It can be referred to as a stochastic process [259]. It is sometimes called "noise" in the time series. When the noise has a Gaussian distribution with zero mean, it is called Gaussian white noise [283].

Figure 2.1 shows the decomposition of the YGT from 1970 until 2023. The decomposition is produced using the forecast R package [139]. The trend component is extracted using a moving average. The seasonal component is computed by averaging through the defined frequency. Since it is a monthly time series, the frequency is 12. The residual component is determined by removing the trend and seasonal components of the time series. The figure also shows that the temperature has a rising trend. The seasonal component has a major sinusoidal yearly (with some spikes). The mean of residual equals zero, and its variance is limited, which seems to be a white noise [283].

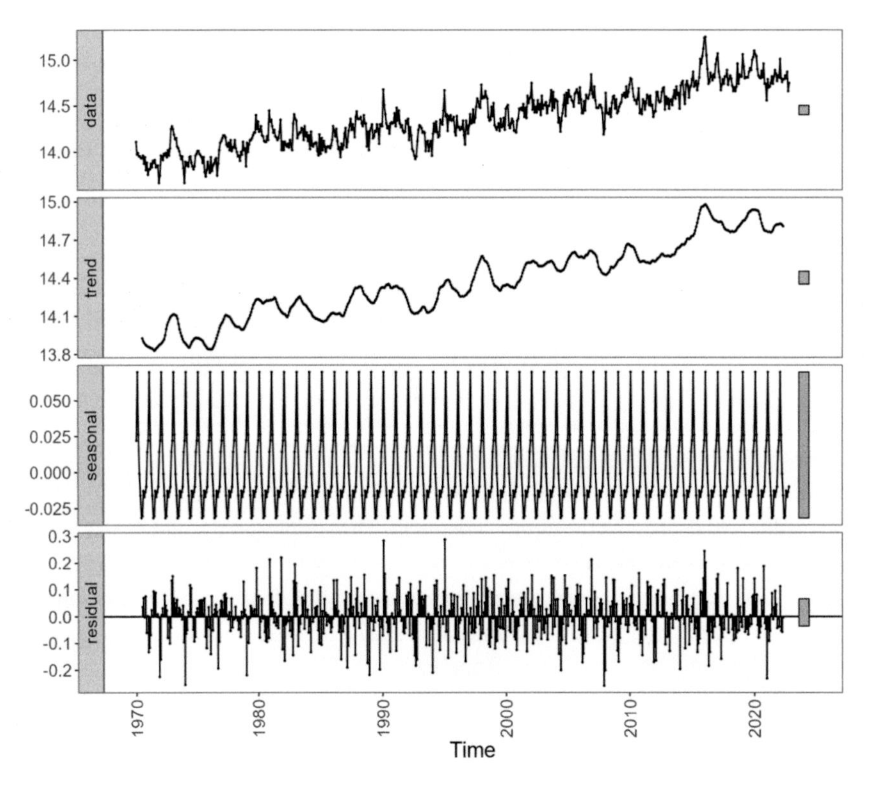

Fig. 2.1 YGT decomposed using R package forecast [139], showing the time series, its trend, seasonal, and residual components

2.2 Stationarity

Most methods applied for time series analysis assume that observations of a time series have a level of regularity over time, which is addressed with the concept of stationarity [119]. A widely adopted definition of stationarity establishes that given a time series X [119]:

i The mean function, $E(X) = \mu$, is constant and does not depend on time.

ii The variance function, $VAR(X) = E\left((X - \mu)^2\right) = \sigma^2$, is constant and does not depend on time.

iii The autocovariance function, $\gamma(X_s, X_t)$, between the time-shifted time series X_t and X_s depends only on the difference $|s - t|$.

In a stationary time series, the statistical properties of the mean, variance, and covariance remain constant over time [308]. These constraints are important since they enable statistical inference based on any sampled time series subset [129]. It is a time series where the data is unaffected by external factors such as trends, seasonality, or other patterns. Figure 2.2a shows an example of a stationary time series, where one may observe mean and variance functions independent of time.

A time series that does not follow a constraint that a stationary process imposes is considered nonstationary. It may manifest in many ways but implies that its mean or variance functions depend on time t [247]. The changes are often due to deterministic trends, structural breaks, level shifts, or changes in variance (a condition known as heteroscedasticity). They can also be due to the presence of unit roots [129].

A trended time series is the simplest form of nonstationarity, with stationary behavior around a deterministic trend [247]. This trend shifts the mean of a time series, causing it to increase or decrease over time. The deviations of a systematic trend may be a stationary variable, known as a detrended variable, which may be analyzed instead of the original time series. In that case, usual stationary models are applicable [129]. A time series that has this behavior is called trend stationary. Figure 2.2b is an example of a trend stationary time series.

Structural breaks may also cause nonstationarity to occur at specific points in time, usually due to environment changes [247]. They may eventually result in level shifts in a time series, which cause the mean function to differ in separate time series intervals. In that case, a time series can be partitioned to separately analyze each data portion with different statistical properties, provided that the timing of a structural break is known [129]. Figure 2.2c shows an example of a level stationary time series where one can see the level shifts of the mean function.

Another cause of nonstationarity is the change in variance over time, known as heteroscedasticity [247]. Heteroscedasticity arises from environmental changes that increase or decrease the volatility of time series observations over time. The time series that has this condition is called heteroscedasticity. An example of a heteroscedastic time series is shown in Fig. 2.2d, where different variance properties on the first and last portions of the series are easily observable.

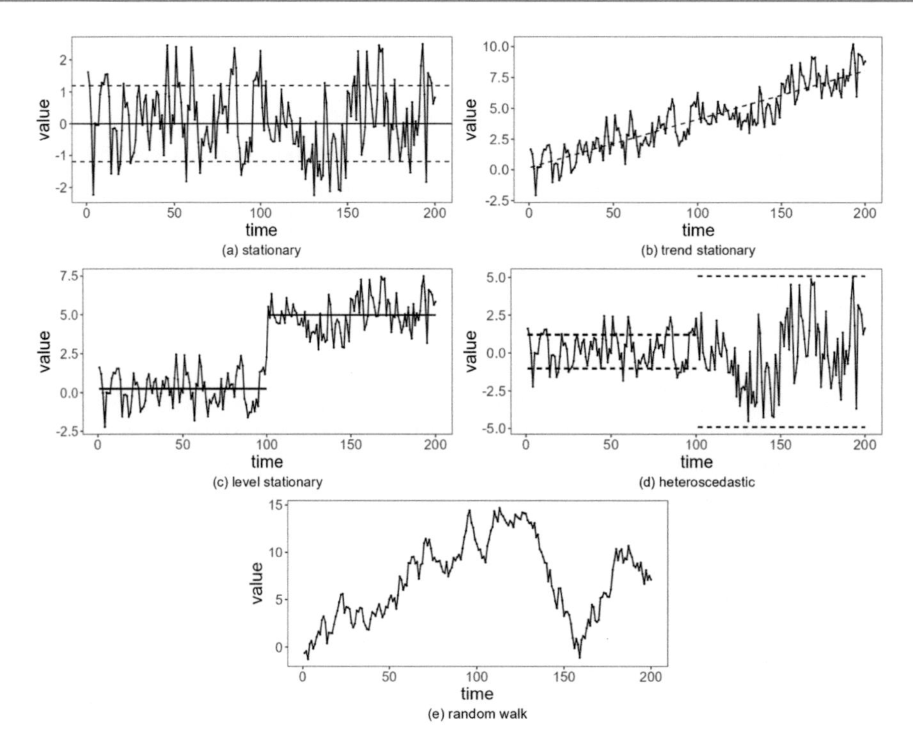

Fig. 2.2 Examples of time series (adapted from Salles et al. [247]): stationary (**a**); trend stationary (**b**); level stationary (**c**); heteroscedastic (**d**); and difference stationary (**e**). Except for **a**, all examples are related to nonstationary time series. The solid and dashed black lines indicate the time series mean and variance functions

An important nonstationarity often observed is caused by a unit root in the characteristic polynomial of a time series model [247]. Without a unit root, time series observations tend to fluctuate around deterministic components such as a mean or a trend. Conversely, observations do not revert to a historical level when a unit root is present and may wander in any direction. The presence of a unit root implies that the time series suffers from the influence of long-run components or stochastic trends [129]. In that case, removing a stochastic trend by applying a process called differencing is often helpful to coerce such time series to stationarity. Therefore, nonstationary time series that have unit roots are also known as difference stationary [45]. Figure 2.2e represents the so-called random walk model [259].

The mean and variance evaluation for stationary time series are relatively intuitive. The test for autocovariance can be computed using Autocorrelation Function (ACF) and Partial Autocorrelation Function (PACF) of the residuals of an First-Order Autoregressive Model (AR(1)) model along with the corresponding confidence intervals for a pair of compared lags [119]. Such discussion is outside the scope of this book, but it can be visually explored by plotting ACF. The ACF measures the correlation between a time series and its lagged values, i.e., measures how much observations in a time series are related to their previous observations. Whenever ACF is above the dashed threshold, that particular lag is related to the current value. While working with stationary time series, as lags increase, ACF rapidly decreases to zero. Conversely, in nonstationary time series, such convergence goes slowly or may not occur. Figure 2.3a–e shows a visual example of ACF for the examples of Fig. 2.2 (stationary, trend stationary, level stationary, heteroscedastic, nonstationary), whereas Fig. 2.3f shows an example of YGT. For both Fig. 2.2a and d, the ACF rapidly goes to zero. The other examples indicate potential nonstationary time series from the perspective of ACF. The heteroscedasticity of Fig. 2.2d is not captured by ACF.

Well-established statistical tests can be used in time series to test stationarity. The Augmented Dickey–Fuller and Phillips–Perron Tests investigate whether a time-ordered set of observations contains a unit root and is, therefore, nonstationary [93]. These tests have the null hypothesis of nonstationary time series. Suppose the p-value is lower than a predefined significance level (5%). In that case, it goes to the alternative hypothesis of stationary time series. Besides, the Breusch–Pagan Test evaluates a linear regression model of the time series to check for heteroscedasticity [119]. Since none of these tests checks all aspects of nonstationarity, they are usually applied in conjunction to confirm the stationarity of a time

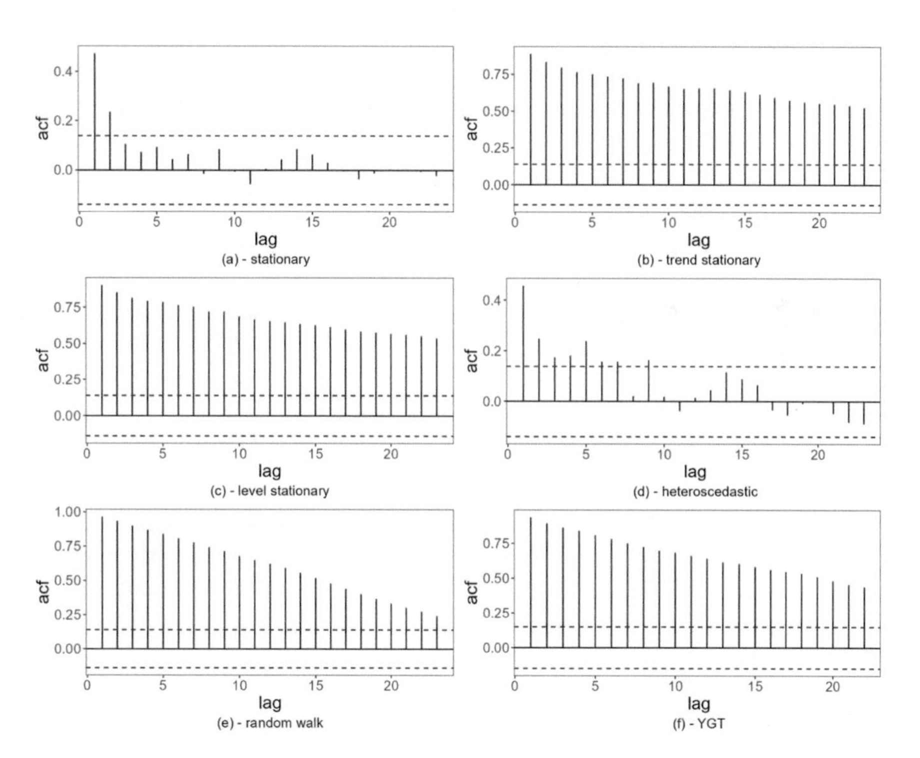

Fig. 2.3 Examples of the ACF for stationary time series of Fig. 2.2a (**a**), 2.2b (**b**), 2.2c (**c**), 2.2d (**d**), and 2.2e (**e**) and the ACF of YGT (**f**)

Table 2.1 Tests for nonstationarity time series. Marked values correspond to the nonstationarity of the time series detected by the statistical test

Time series	Augmented Dickey–Fuller test	Phillips–Perron test	Breusch–Pagan test
Stationary			
Trend stationary			
Level stationary	X		
Heteroscedastic			X
Difference stationary	X	X	X
YGT			X

series. Table 2.1 evaluates the time series of Fig. 2.2 and the YGT. All series marked with X are considered nonstationary according to the respective test. Since the trend stationary time series is a simple case of nonstationarity, none of these tests consider it nonstationary. The level stationary case is detected by the Augmented Dickey–Fuller Test. The heteroscedasticity is only detected by the Breusch–Pagan Test. All three applied tests detect the difference stationary. The YGT is only marked as nonstationary by the Breusch–Pagan Test.

The dependence of a time series on past data may occur frequently by multiples of some underlying seasonal lag S. In that case, a time series exhibits periodic components. Therefore, its statistical properties, such as mean and variance, may periodically change, creating a dependence on time t, thus making seasonality another form of nonstationarity often found in time series [308].

If not properly addressed, any nonstationarity can have a severe impact on time series modeling, leading to fallacious statistical inferences and bad or unexpected results. The next section categorizes some relevant data preprocessing for the event detection subject. Many of them are prepared to address nonstationary time series.

2.3 Data Preprocessing

Data preprocessing is an important activity in any application of data analytics. It commonly encompasses more than 60% of the data mining process. Data cleaning, feature selection, sampling, outlier removal, normalization, and data transformation are commonly performed during preprocessing [235]. The main goal of data preprocessing is to guarantee the quality of the data before it is fed to any learning algorithm [108]. This context encompasses problems of classification (prediction of discrete data) and regression (modeling or prediction of continuous data) [125]. This chapter focuses on transformation methods that target the regression problem, which are useful for modeling and aiding the actual regression of the time series. These tasks are relevant during time series event detection. While there are many tasks involved during the time series preprocessing, it is possible to summarize the most common ones as follows:

- Data cleaning: It is the basis for data analytics activities [298]. It involves handling missing values, outliers, and errors in the data. Missing values can be filled in using techniques such as interpolation or imputation. Some outliers can be detected using statistical methods. They are commonly removed or smoothed. However, outliers in the scope of this book are generally addressed as anomalies and are the data wanted to be discovered. They are presented in Chap. 3.
- Temporal aggregation: Depending on the frequency of the data, it may need to be resampled to a lower frequency to match the desired analysis time frame. Temporal aggregation is covered in Sect. 2.3.1.
- Data transformation: It involves creating new features or variables from the existing data to capture additional information or relationships. It includes trend components extraction, variance stabilization, detrending and differencing, and decomposition [259], which are covered in Sects. 2.3.2, 2.3.3, 2.3.4, and 2.3.5.
- Sliding window: The basic idea is to divide a time series into smaller subsequences, which provide ways to perform time series analysis or modeling on each subsequence separately [209]. Sliding windows are covered in Sect. 2.3.6.
- Data normalization: It involves scaling the data to a common range or unit to be compared or combined with other data or to support ML methods. Normalization techniques include Standard Score Normalization (Z-Score), Min-Max, and log transformation [125]. Data normalization is covered in Sect. 2.3.7.
- Data splitting: Partitioning the data into training, validation, and testing sets is important to enabling model build and evaluation [139]. Data splitting is covered in Sect. 2.3.8.

2.3.1 Temporal Aggregation

Temporal aggregation refers to summarizing time series from higher to lower frequencies, as shown in Fig. 2.4. It can be applied to higher (such as minutes, hours, and parts of the day) or lower frequency time series (days, weeks, months, quarters, and years). The purpose of temporal aggregation is to reduce the level of detail in the data, which can help to simplify analysis and visualization, as well as to remove noise or fluctuations that may be present at the finer time scales.

Temporal aggregation is applied in many fields, such as finance, economics, and environmental monitoring. Data is often collected at high frequencies but may need to be analyzed or reported at lower frequencies. However, temporal aggregation can also result in loss of information and may affect the accuracy of statistical models or forecasts [205].

Let X be a time series taken at equidistant time intervals, for example, the YGT. One may often be interested in studying not monthly (which may be interfered with by season

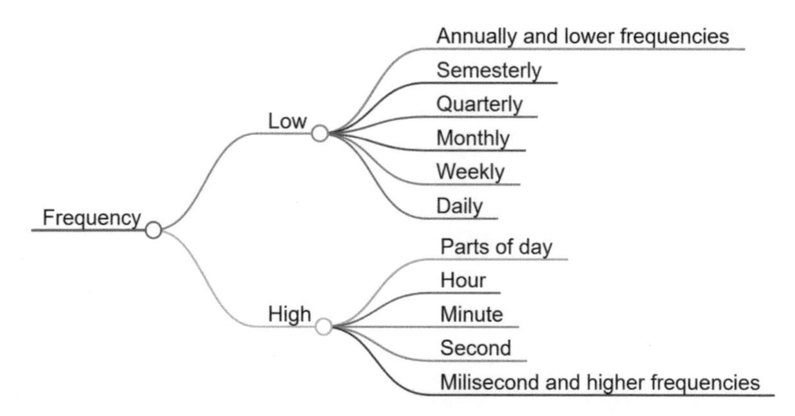

Fig. 2.4 Temporal aggregation according to frequency level

effects) but every year, which are temporal aggregations from monthly data [278]. Taking x_i as observations of X, a non-overlapping aggregated m-period time series Y_m is defined in Eq. 2.2, where y_j are observations of Y_m is the aggregated time series and m corresponds to the m-period aggregates of X, i.e., the order of aggregation. Thus, in our previous example, t represents the time unit of the month, and m is 12 [295].

$$y_j = \sum_{t=m(j-1)+1}^{m\cdot j} \frac{x_t}{m} \tag{2.2}$$

Figure 2.5a shows MGT, whereas Fig. 2.5b shows YGT. As observed, YGT preserves the average observation of MGT each year, removing seasonal components associated with monthly values.

2.3.2 Trend Component Extraction

The Linear Regression is the simplest and most rigid method for extracting trend components from a time series. It can be described by Eq. 2.3, where $\hat{x}_t$ is the trend at time t. Parameters α and β are obtained from mean square error adjustment, and ω_t is the residual from this Linear Regression adjustment [145]. Figure 2.6a shows an example of Linear Regression applied to YGT.

$$\hat{x}_t = \alpha t + \beta + \omega_t \tag{2.3}$$

The Moving Average Smoother is widely used, in particular, in finance and econometrics, to highlight seasonality and long-term trends in a time series [259]. Moving Average Smoother can detect the evolving trend of a time series by minimizing random noise [209] and can also be used for seasonal adjustment of a time series. One of the simplest forms of

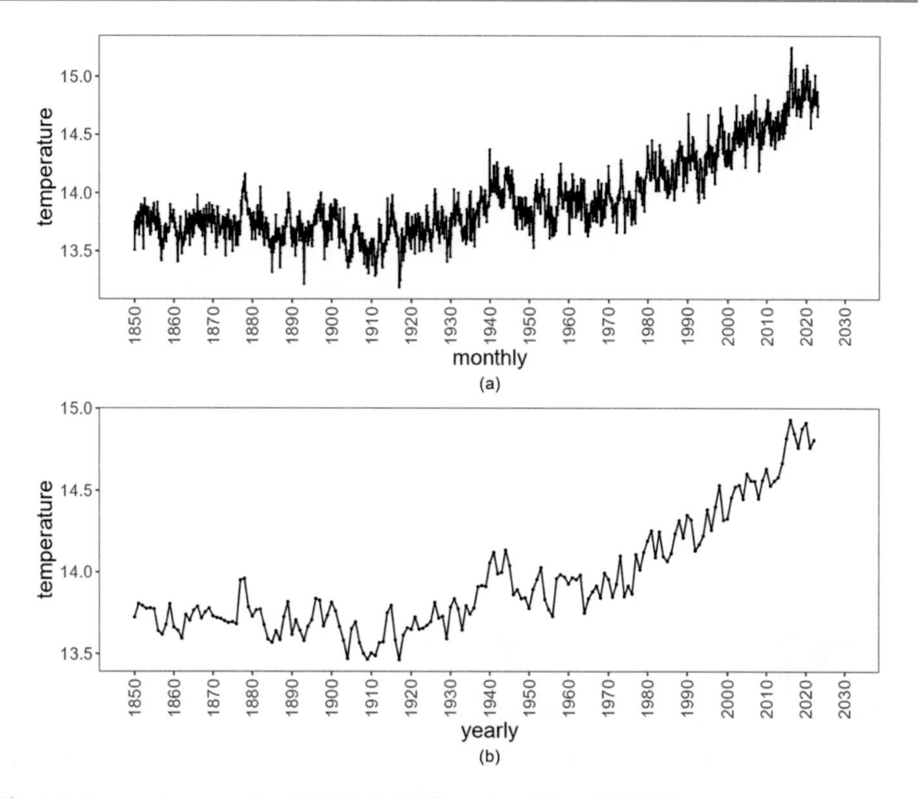

Fig. 2.5 Temporal aggregation: MGT (**a**); YGT computed from MGT (**b**)

the Moving Average Smoother is represented by Eq. 2.4, where k is the order of the moving average. Figure 2.6b shows an example of Moving Average Smoother applied to YGT.

$$\hat{x}_t = \frac{1}{k} \sum_{i=t}^{t+k-1} x_i, \quad 1 \leq t \leq n - k + 1 \tag{2.4}$$

There are more advanced techniques for extracting the underlying trend components of time series, some of which we present in Sect. 2.3.5. Other techniques involve segmenting the time series when changes in the trend occur, as presented in Chap. 4.

2.3.3 Variance Stabilization

Variance stabilization is a technique used in time series analysis to transform data with non-constant variance into data with constant variance. In many time series, the variance changes, making it difficult to analyze the data using traditional statistical methods. The

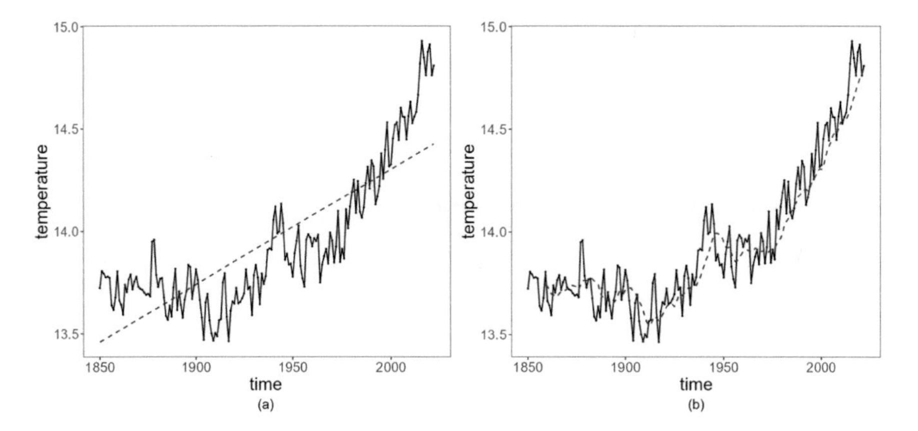

Fig. 2.6 Extracting trend components from YGT: Linear Regression (**a**); Moving Average Smoother (**b**)

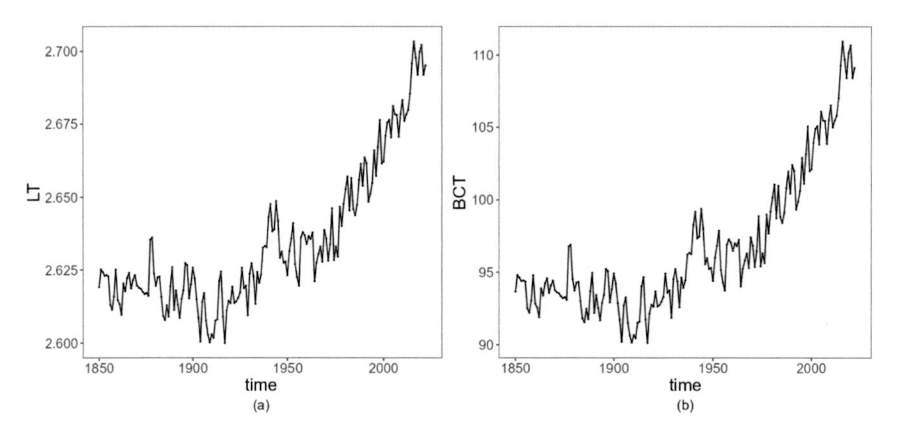

Fig. 2.7 Variance stabilization applied to YGT: Logarithmic Transform (**a**); BCT (**b**)

most used methods for addressing variance stabilization are the Logarithmic Transform and BCT [78].

The Logarithmic Transform tends to suppress fluctuations that occur over portions of a time series that present higher values [259], and its simple formulation can be seen in Eq. 2.5, where $\hat{x}_t$ is the transformed version of the original time series value x_t. Often in macroeconomics, a time series is a natural log-transformed (setting $\mathrm{b} = e$) to minimize effects of nonstationarity and heteroscedasticity (non-constant variability [119]) and induce symmetry and normality. The estimated coefficients of a logged time series can also be interpreted as elasticity. These advantages make natural logs one of the most applied Logarithmic Transform. Figure 2.7a shows an example of Logarithmic Transform applied to YGT.

$$\hat{x}_t = \log_{\mathrm{b}} x_t \tag{2.5}$$

The BCT, as in Eq. 2.6, provides a more objective method for determining a suitable power transformation of a time series [313]. However, it requires calculating the numeric argument λ related to the data distribution function, which is not always known and can only be applied to positive valued data [78]. Figure 2.7b shows an example of BCT applied to YGT.

$$\hat{x}_t = \begin{cases} \left(x_t{}^{\lambda} - 1\right)/\lambda, & \lambda \neq 0, \\ \log x_t, & \lambda = 0 \end{cases} \tag{2.6}$$

2.3.4 Detrending and Differencing

The transformations of Detrending (or trend removal) and Simple Differencing are widely used in combination with time series modeling methods. The Detrending transformation involves determining and removing an inherent trend observed in a time series [259]. The observed trend is estimated and defined by a deterministic functional form, which may be represented as a fixed component in a time series model [301]. A general Detrending transformation is shown in Eq. 2.7, where η_t is the determined trend component and, in this case, $\hat{x}_t$ represents the variability series, i.e., the residue of the time series after trend removal. Figure 2.8a shows an example of detrending using linear regression.

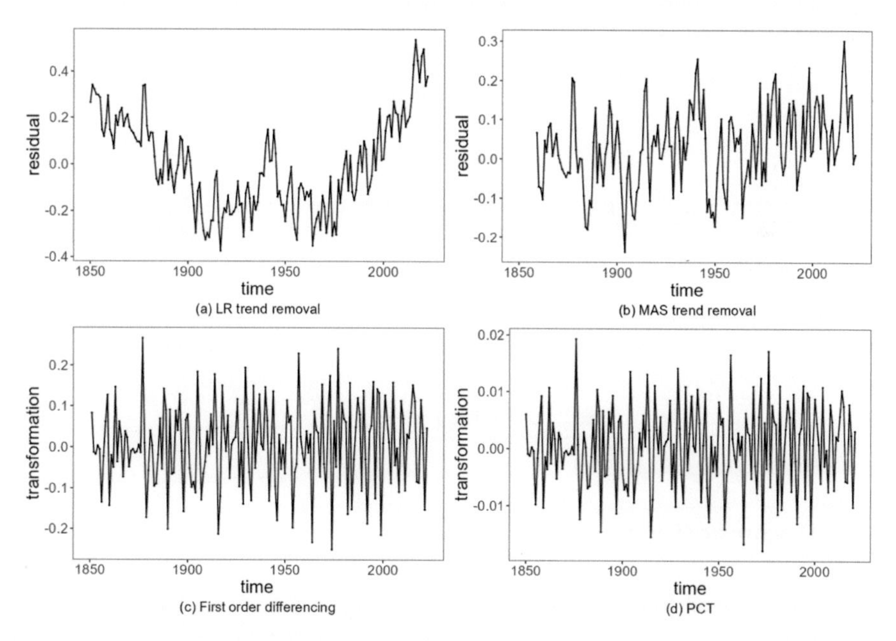

Fig. 2.8 Detrending versus differencing: Detrending of YGT using Linear Regression (**a**); Detrending of YGT using Moving Average Smoother (MAS) (**b**); Simple Differencing of YGT (**c**); Percentage Change Transform (PCT) of YGT (**d**)

$$\hat{x}_t = x_t - \eta_t \tag{2.7}$$

This method assumes that the deterministic trend is always appropriate, which is frequently not the case in many applications, in particular, involving nonstationary processes [78]. Moving Average Smoother (MAS) is usually adopted to overcome this drawback as a way to smooth out random fluctuations in data and identify underlying trends. Equation 2.8 computes the moving average η_t for time series X. Parameter k is the order of the average, as the number of previous observations used to compute the moving average and introduces inertia to the MAS. The higher the k, the more observations are needed to follow a change in the time series trend. The first $k - 1$ points for η_t are not defined due to the absence of observations to compute the MAS. An example of MAS for the YGT is shown in Fig. 2.8b.

$$\eta_t = \frac{\sum_{i=0}^{k-1} x_{t-i}}{k} \tag{2.8}$$

The Simple Differencing method brings some advantages compared to Detrending since it does not require a parameter estimation process and is capable of generating a stationary time series when having stationary behavior around a deterministic or stochastic trend [259]. A first Simple Differencing transformation, written as in Eq. 2.9, can eliminate a linear trend, a second Simple Differencing eliminates a quadratic trend, and so on. For Simple Differencing of higher order, the backshift operator (B) is used, as seen in Eq. 2.10, where ∇^d is the d-th differencing and the operator $(1 - B)^d$ is adapted for higher orders of d [259]. An example of Simple Differencing for the YGT is shown in Fig. 2.8c.

$$\hat{x}_t = \nabla x_t = x_t - x_{t-1} \tag{2.9}$$

$$\nabla^d = (1 - B)^d, \; B^k x_t = x_{t-k} \tag{2.10}$$

The Percentage Change Transform can be an alternative to Simple Differencing, assuming considerable stability in the relative percentage of change between the two following observations of a time series [78]. In that case, Eq. 2.11 is true if the percentage change is restricted to an interval $[-100\rho, 100\rho]$, with $0 \leq |\rho| < 1$ being a small acceptable threshold. Figure 2.8d shows an example of Percentage Change Transform applied to YGT.

$$\hat{x}_t \approx \log\left(\frac{x_t}{x_{t-1}}\right) \tag{2.11}$$

2.3.5 Decomposition

Decomposition-based transformations originate in signal processing techniques such as frequency-domain analysis [181], decomposing a time series into components (signals) having different scales (frequencies), and targeting to capture the intrinsic dynamics of a time series [197]. For example, a time series can be decomposed into short-term (high-frequency), seasonal, and long-term (low-frequency) components. An advantage is that explaining only a few signal components is generally simpler and more physically meaningful than a collection of estimated model parameters [259]. Furthermore, the derived components may be easier to model, which can simplify the time series modeling [91]. Decomposition may be achieved in a frequency-only or a time–frequency domain, which enables the preservation of information regarding localized changes. Special cases of time series decomposition are based on moving average iterations or pattern mapping (i.e., deriving patterns of a time series to simplify its modeling and filter trends and long-term variations [91]). The following subsections describe some of the most frequently used decomposition-based transformation methods.

2.3.5.1 Frequency-Domain Decomposition

Time series analysis in the frequency domain is often based on Fourier Transform [149], which creates a frequency-based representation (a spectrum) of a time series in terms of Fourier basis functions. The Fourier Transform of a time series X, in this case, represented as a function of time x_t, can be formulated as in Eq. 2.12, where $F(\xi)$ represents the Fourier spectrum, ξ is a frequency component, and j is the imaginary number $j = \sqrt{-1}$ [181].

$$F(\xi) = \int_{-\infty}^{+\infty} x_t e^{-j\xi t} dt, \quad e^{-j\xi t} = \cos \xi t - j \sin \xi t \tag{2.12}$$

Fast Fourier Transform (FFT) is often used to compute a sequence's Discrete Fourier Transform as it is much faster than computing the Discrete Fourier Transform using the standard Fourier Transform method [77]. It is, in particular, useful in time series analysis, where it can be used to identify periodic patterns in the data by decomposing the time series into harmonics. An example of decomposition using FFT, applied to YGT, is shown in Fig. 2.9.

Figure 2.9a shows that the YGT contains 90 harmonics, which, combined, can recover the entire YGT with imperceptible error. However, it is interesting to use a small number of harmonics to capture the main trend, using periodogram [51]. A technique is to establish a frequency threshold, for example, by visually inspecting the periodogram and identifying the frequency values where the spectral density drops off sharply or appears to level off. Another technique is to evaluate the appropriate number of harmonics by model selection criteria, such as Akaike Information Criterion (AIC) or Bayesian Information Criterion (BIC) [139]. Figure 2.9b shows the periodogram for YGT. A frequency threshold is set at 0.06. It led to the adoption of the major nine harmonics.

These nine harmonics can capture major trends and seasonal components of YGT as shown in Fig. 2.9c. However, this reduced number of harmonics cannot completely fit the YGT in the borders. In general, when using a small number of harmonics, the FFT provides a better approximation of the time series in the middle of the series. A common practice to address these border effects is to avoid the initial and final observations [47]. Dashed lines in Fig. 2.9c present that FFT is better adjusted after the first and before the last nine

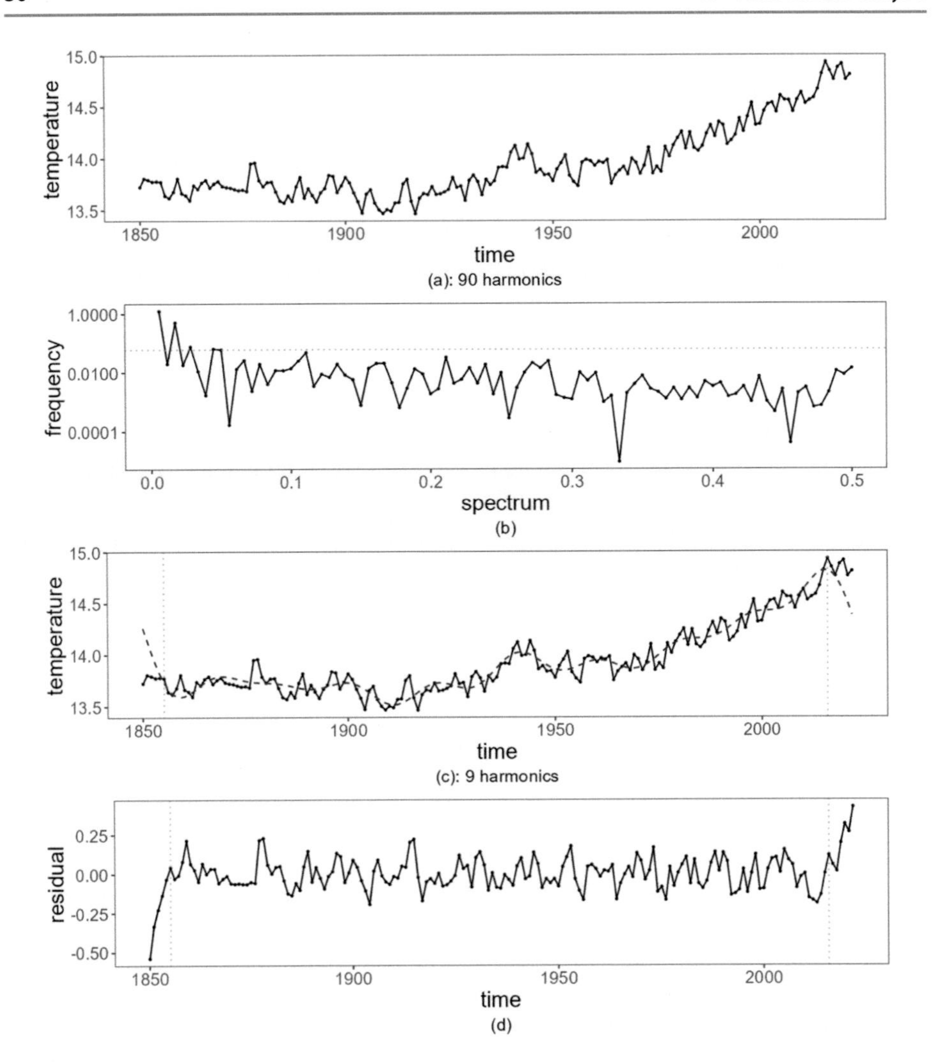

Fig. 2.9 The decomposition of YGT using FFT: FFT with all 90 harmonics (**a**); periodogram of YGT filtered frequency at 0.06 leading to 9 harmonics (**b**); FFT with all nine harmonics (**c**); residual of YGT for FFT with nine harmonics (**d**)

observations. Finally, Fig. 2.9d shows the residual of YGT and the FFT with nine harmonics. Again, the residuals are more significant at the beginning and end of the series.

2.3.5.2 Time–Frequency Decomposition

Wavelets are finite basis functions localized in both time and frequency. The Wavelet Transform decomposes a time series (signal) by correlating it with a family of wavelets, providing an extremely flexible time–frequency representation [164]. It decomposes a time series X, which is again regarded as a function of time x_t, into the wavelet series $\hat{x}_t$ in Eq. 2.13 [149]. The component ζ_t and its coefficient b represent the scale part of the wavelet series (responsible for modeling trends and seasonality). In contrast, the component ψ_t and its coefficient c represent the detail part of the wavelet series (corresponding to noise or random deviations) at scale (decomposition level) l and position k. L is the defined maximum decomposition level.

$$\hat{x}_t = \sum_{k=1}^{n} b_{l,k}\zeta_{l,k_t} + \sum_{l=1}^{L}\sum_{k=1}^{n} c_{l,k}\psi_{l,k_t} \tag{2.13}$$

The Wavelet Transform can be implemented as a DWT [75]. The DWT uses a filter to decompose a signal into different frequency components and works by convolving the signal with two sets of filters: Low-Pass Filter and High-Pass Filter. The Low-Pass Filter and High-Pass Filter filters are designed using the scaling and wavelet functions of a given wavelet family [222].

Figure 2.10 applies the Haar Filter for the YGT. Figure 2.10a shows the $\zeta(t)$ component of DWT (trend and seasonal), whereas Fig. 2.10b shows the $\psi(t)$ component (residual).

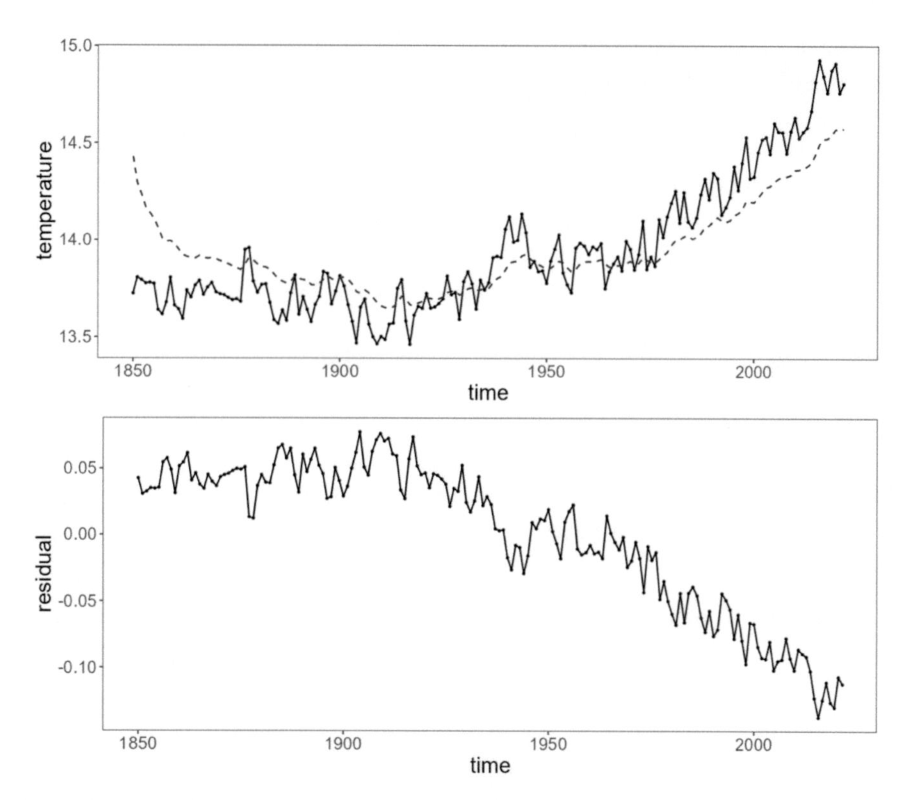

Fig. 2.10 Decomposition of YGT using DWT with Haar Filter filter: $\zeta(t)$ component of DWT (trend and seasonal) (**a**); the $\psi(t)$ component (residual) (**b**)

Similarly to the Discrete Fourier Transform, the adjustment is better in the middle of the time series and faces some adjustment problems at the borders.

Since finite wavelets are irregular and asymmetric [149], the Wavelet Transform is, in particular, suited for analysis of nonlinear, noisy, and nonstationary time series, with rapidly changing observations [24]. If the wavelet transform is applied to a nonstationary time series, the resulting decomposed series behaves better than the original series and is more easily and accurately modeled, even by simple linear models like ARIMA. Predictions of the original time series with increased accuracy can be obtained by applying the inverse Wavelet Transform to the predictions of the decomposed series [75].

Similar to the Wavelet Transform, the Empirical Mode Decomposition (EMD) is a method for nonlinear and nonstationary time series decomposition, generating a time–frequency representation of the series [290]. The basic principles of EMD are described as follows [101]:

1. Identify all extremes of a time series X.
2. Interpolate between minima (resp. maxima), ending up with two *envelopes*: $emin(X)$ and $emax(X)$.
3. Compute the average time series $m(X) = \frac{emin(X)+emax(X)}{2}$.
4. Extract the detail $d(X) = x(X) - m(X)$.
5. Iterate on the residual $m(X)$.

EMD is useful for decomposing a time series. When the detail time series $d(X)$ has zero mean, it can be considered an Intrinsic Mode Function (IMF). IMFs are more stable components that can be more easily modeled [269]. The last residual is considered a major trend component of the time series. However, some of the last IMF identified in the EMD can also be used to model the trend components. The first IMF is associated with the time series residual components. Figure 2.11a uses the residual of IMFs, capturing the trend of YGT, while Fig. 2.11b shows the first IMF, which stands for the time series residual for the YGT.

EMD does not depend on predetermined (wavelet) functions since its basis functions are derived directly from the time series, thus making it adaptive and completely data-driven [290]. Conversely, EMD is recursive and does not perform well when separating time series components with similar frequencies.

2.3.6 Sliding Windows

Sliding windows are widely used in time series analysis to enable local processing [210]. They provide support for ML methods [294] and can be used to support memory management of the time series [183], i.e., the ability to forget past data.

The simplest way to formalize a sliding window is by building it from subsequences. A subsequence of size p obtained from a time series X that ends at the i position can be

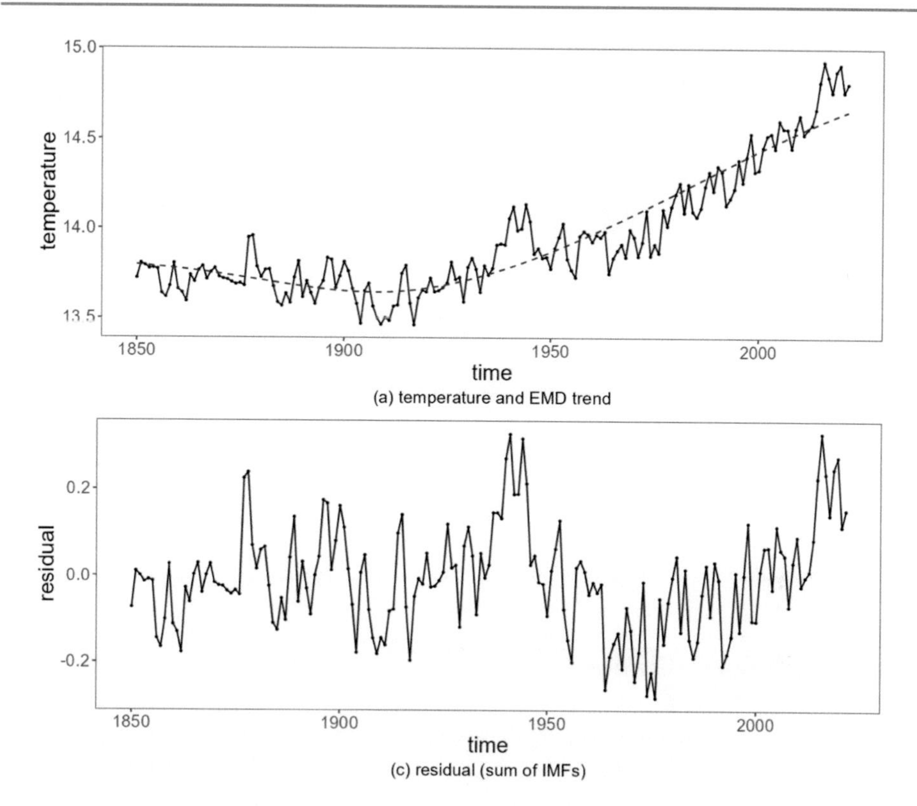

Fig. 2.11 Decomposition of YGT using EMD: EMD residual components, which stands for the time series trend (**a**); EMD only the first IMF, which stands for the time series residual (**b**)

represented by $seq_{i,p}(X)$, which is a continuous sequence of values $< x_{i-(p-1)}, x_{i-(p-2)}, \ldots, x_i >$, where $|seq_{i,p}(X)| = p$ and $p \leq i \leq |X|$. Subsequences enable the analysis of data samples to evaluate local properties [71].

A sliding window explores all possible subsequences of a time series [209]. A sliding window of size p for a time series X is a function $sw_p(X)$ that produces a matrix W of size $(|X| - p + 1)$ by p. Each line w_i in W is the ith subsequence of size p from X. Given $W = sw_p(X)$, $\forall w_i \in W$, $w_i = seq_{i,p}(X)$ [44].

Figure 2.12 shows the YGT from 2000 until 2023, with both the time series and its colored representation. Each rectangle corresponds to an observation. The colors are associated with the values. Lighter colors correspond to lower temperatures (close to 14.25), while darker colors correspond to higher temperatures (close to 15). Figure 2.12 also shows a colored example of a sliding window of size 3. Each triple of t, $t-1$, and $t-2$ corresponds to a separate sliding window with three rectangles aligned vertically. It starts in the third observation with its first complete sliding window. Figure 2.12 lists observations of the first five sliding windows.

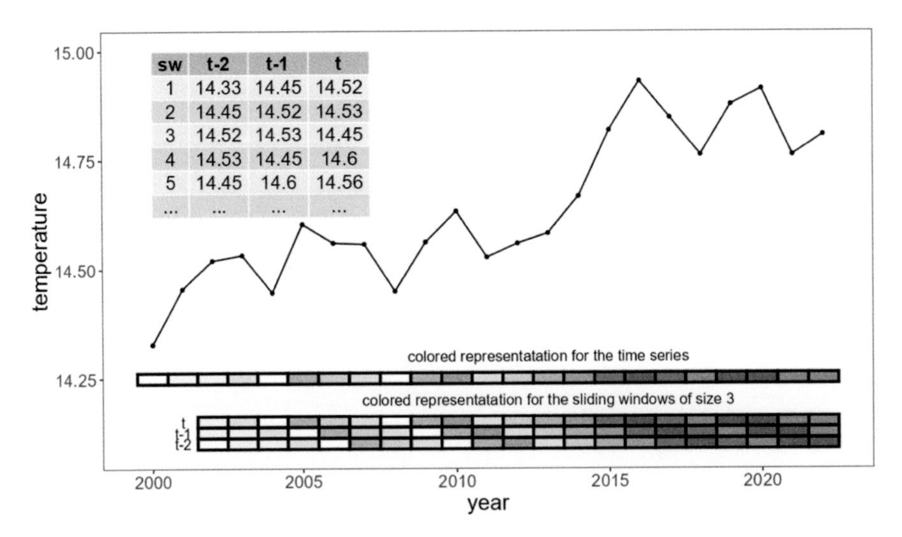

Fig. 2.12 Subsequences and sliding windows of time series (adapted from Ogasawara et al. [209])

2.3.7 Data Normalization

Data normalization transforms data into a regular scale to make it more consistent and easier to compare. Normalization of time series is important because the values in the series may vary widely over time. It can be applied to the entire time series or sequences of the series to reduce the impact of outliers, make it easier to compare trends over time, and ease the use of certain analysis methods, such as ML algorithms [210].

The normalization techniques during data transformation include Min-Max, Z-Score, and decimal scaling. The Min-Max method normalizes the values of a time series X according to its minimum and maximum values, converting a value x_i of X to $\hat{x_i}$ in the range $[low, high]$ (see Eq. 2.14). It is common to set low equal to 0 and $high$ equal to 1.

$$\hat{x_i} = (high - low) \cdot \frac{x_i - min(X)}{max(X) - min(X)} + low \tag{2.14}$$

The decimal scaling normalization moves the decimal point of the values of a time series X according to its maximum absolute value. Hence, a value a of x_i is normalized to $\hat{x_i}$ as described in Eq. 2.15, where d is the smallest integer such that $max(|\hat{x_i}|) < 1$.

$$\hat{x_i} = \frac{x_i}{10^d} \tag{2.15}$$

In the Z-Score normalization, the values of a time series X are normalized according to their mean ($\mu(X)$) and standard deviation ($\sigma(X)$). A value x_i of X is normalized to $\hat{x_i}$ as described in Eq. 2.16.

$$\hat{x}_i = \frac{x_i - \mu(X)}{\sigma(X)} \tag{2.16}$$

These techniques are useful in stationary environments but are limited to nonstationarity. They may be effective if differentiation is applied before normalization. Data normalization can also be applied after sliding windows transformation [130]. In this case, instead of considering the complete time series for normalization, it normalizes each window, considering only its statistical properties. The rationale behind this technique is that decisions are usually based on recent data. The sliding windows technique can always normalize data in the desired range, yet assuming that the time series volatility is uniform, which is not true in many phenomena [15]. To overcome sliding window limitations, we have proposed Adaptive Normalization to address nonstationarity, especially to support ML methods [209]. The general process of Adaptive Normalization encompasses the following steps:

1. Compute the moving average for each sliding window.
2. Compute the residual of each sliding window with respect to its moving average.
3. Compute the distribution of residuals for all sliding windows and remove all windows with outliers according to box plot criteria [125].
4. Normalize all sliding windows using this Min-Max extracted from the cleaned distribution of residuals of all sliding windows.

Figure 2.13 shows the effects of using different normalization techniques. Figure 2.13a shows the YGT marking a subsequence of size five between 1924 and 1928, while Fig. 2.13b shows the normalization of that sequence using global Min-Max. Since observations are close to

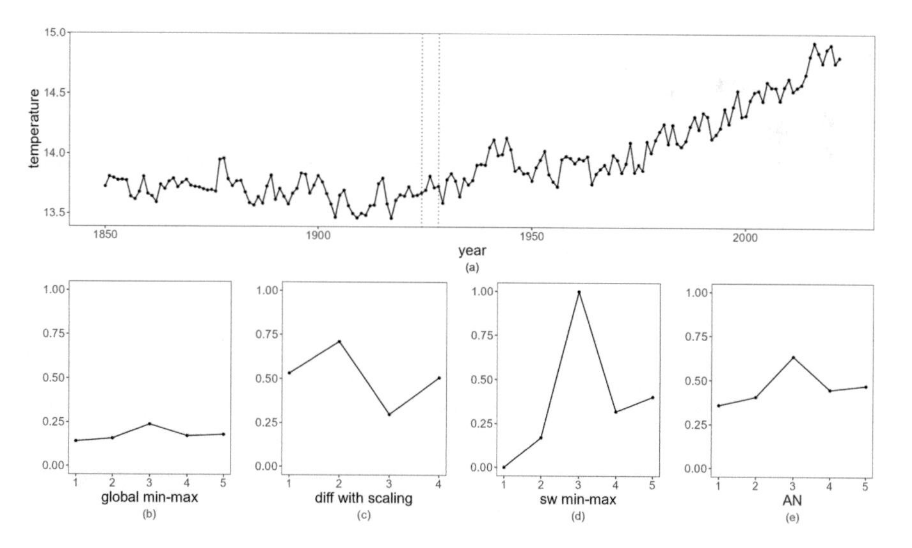

Fig. 2.13 Time series normalization

the lowest values of YGT, the normalization values are close to 0. The main problem with using the global Min-Max is that the minimum and maximum values of the out-of-sample dataset may lead to values higher than one and lower than zero, which may not be supported by ML models [209].

Figure 2.13c shows the effect of using differentiation before data normalization. The limitation of global Min-Max is not a problem for this technique. However, the differentiated time series can differ considerably from the original one. For example, values could decrease due to a decrease in the growth rate, even though the time series increases.

Figure 2.13d shows the Min-Max applied to each sliding window, where all sliding windows distribute values between 0 and 1. The drawback is that this normalization may stretch values with small changes in the same way as values with big changes. Such a characteristic might inhibit the performance of ML models. Figure 2.13e shows the Adaptive Normalization, where differentiation is first applied to the moving average and then Min-Max using the minimum and maximum values for all sliding windows. Thus, Adaptive Normalization can preserve the original time series shape and spread values between 0 and 1, considering the ratio of values changes inside the sliding window.

2.3.8 Data Splitting

Splitting the data involves dividing the available time series into sets for training, validation, and testing purposes. It allows evaluating the model's accuracy and prevents overfitting, which occurs when it is too complex and captures noise or random fluctuations in the training data, leading to a poor generalization of new observations [125]. The most common technique is to use sliding windows [139], which involves selecting a fixed interval for training the model and then using the trained model to predict subsequent observations. The process is repeated for the entire time series, with the sliding window moving forward at a fixed interval each time.

In addition to the training data, the validation data is used to tune the model's hyperparameters, such as the number of lagged values to include. In contrast, the testing data is used to evaluate the final performance of the model. A common splitting ratio is $\frac{2}{3}$ for training and $\frac{1}{3}$ for testing. The optimal ratio may vary depending on the size and complexity of the time series [125].

While training and validating or training and testing, a distinct difference exists between traditional cross-validation techniques and time series cross-validation. Time series cross-validation is a technique for evaluating the performance of a time series model by testing it on a sequence of data that is distinct from the training data. It is a variation of the traditional cross-validation method used in ML but designed specifically for time series. In time series cross-validation, the data is divided into consecutive sequences, where each consecutive sequence is used for evaluation (validation or testing). The remaining prior sequences are used as the training set. This process is repeated for each sequence, and the results are averaged to estimate the model's performance. Thus, the key difference between time series cross-validation and traditional cross-validation is that the sequences are selected based on their temporal ordering rather than randomly. Such selection occurs since time series has a temporal structure, where each observation depends on previous observations, and random sampling can introduce bias and unrealistic assumptions [139].

Figure 2.14a shows the YGT from 2000 until 2022, separating 2000 until 2018 for training and separating the last four observations for testing. To ease visualization, we use the colored representation for the time series. As shown in Fig. 2.14b, a one-step-ahead prediction uses the model built from the training set to predict the following observation. The one-step-ahead prediction with time series cross-validation uses the model built during the training set to predict the first observation of testing. The second prediction uses the built model sliding one step as if the first testing observation was known to predict the second. This

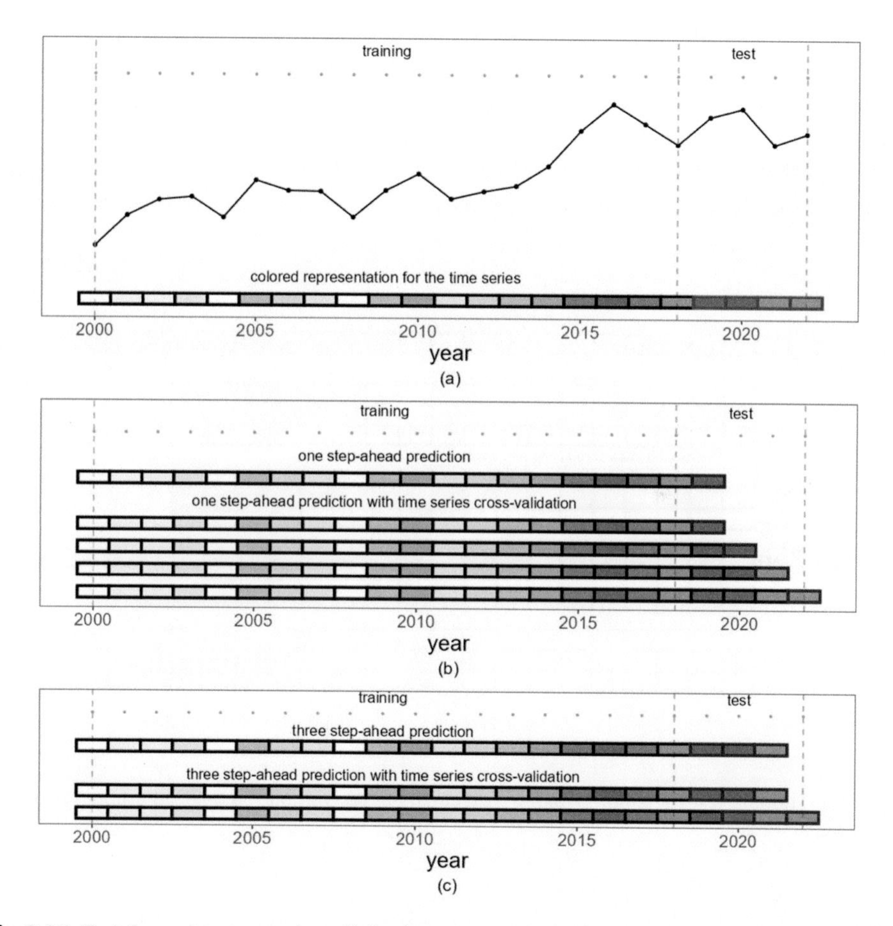

Fig. 2.14 Training and test example: splitting between training and test (**a**); one-step-ahead prediction with time series cross-validation (**b**); three-step-ahead prediction with time series cross-validation (**c**)

process is repeated for all testing observations. The average error of prediction for all testing observations measures the Prediction Performance.

Figure 2.14c shows a three-step-ahead prediction that uses the model built from the training set to predict the following three observations. The average prediction error for the three predictions measures the Prediction Performance. The three-step-ahead prediction with time series cross-validation uses the model built during the training set to predict the first three observations. The second prediction uses the built model sliding one step as if the first testing observation was known to predict the following three observations. This process is repeated until all testing observations are explored. The average error of all predictions (six in this example) measures the Prediction Performance [139].

Finally, we consider the scenario where the time series model is built using a sliding window. This scenario corresponds to the typical scenario of ML. Figure 2.15 depicts the colored series converted to a sliding window. The sliding window size equals three and contains 23 sequences. The sliding window data is also split. The model is built using only the 17 sequences dedicated to training. In a one-step-ahead prediction, the model uses the last sequence before testing to predict the first test observation. Then, the model uses the first testing sequence to predict the second observation, and the process is repeated until the last prediction is made [209]. A similar process also occurs for n-step-ahead prediction using a sliding window, where the step-ahead uses predicted sequences instead of real observations.

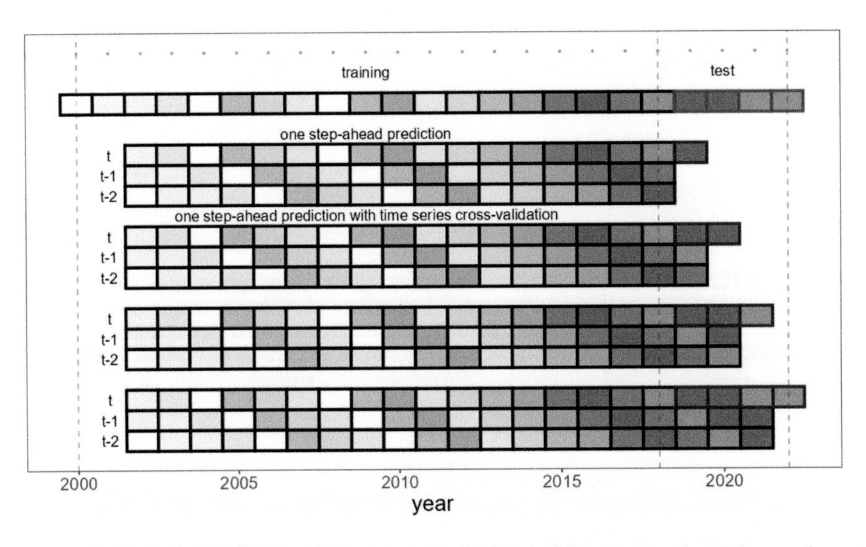

Fig. 2.15 Training test for sliding window size equals three with one-step-ahead time series cross-validation

2.4 Time Series Prediction

Time series prediction encompasses both problems of classification (prediction of discrete data) and regression (prediction of continuous data) [54]. This section focuses on the problem of time series prediction through regression. For simplicity, we refer to prediction and regression interchangeably. The models adopted for time series prediction generally fall into the categories of statistical or ML models [246]. The accuracy of the predictions depends on the quality of the historical data, the appropriateness of the model, and the assumptions made about the underlying processes driving the time series [139]. Figure 2.16 shows a general time series prediction process. It encompasses five main activities. It provides a general framework for predicting a time series based on a particular setup of preprocessing methods and prediction models. They are briefly described here, and some parts are detailed in the following sections.

The first activity in Fig. 2.16 refers to acquiring the time series and performing data preprocessing by applying time series transformation methods, such as those described in earlier sections. It also includes data normalization and sliding windows transformation. These transformations change the time series domain values, and their parameters must be stored to support later detransformation to the original domain. For time series prediction, splitting the time series into a training and test set is also important during Activity 1. All data preprocessing parameters should be computed during training and reapplied from the tune-values of training during the test. The model is built using the training slice and evaluated

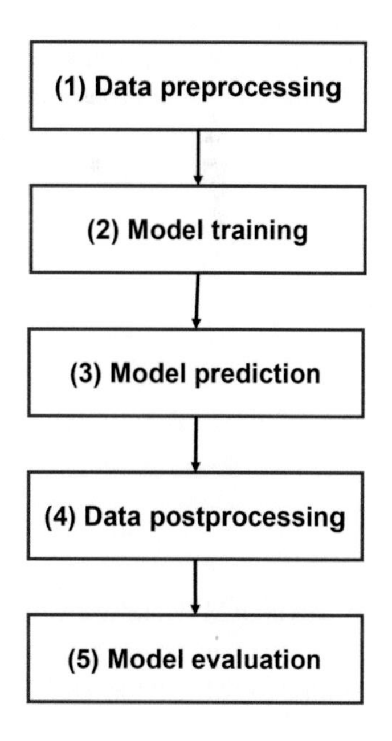

Fig. 2.16 Time series prediction process (adapted from Salles et al. [246])

using the unseen test set. However, when the goal is to adjust a model for the time series, the model does not need to be partitioned into a training and test set.

Activity 2 addresses model training. The prediction methods very often require hyperparameter optimization. In such a case, the training slice is again split into a novel training and validation set. Alternative models exploring hyperparameter values are built using the novel training and evaluated using the validation set. Once hyperparameters are fixed, a single model is built using the entire training time series. Then, the model is available for use. Activity 3 refers to the model prediction. Since the predicted values are not in the time series domain, they cannot be directly evaluated, so data postprocessing is needed. Activity 4 corresponds to the postprocessing of predictions, reversing transformations applied to the time series in Activity 1. Finally, Activity 5 evaluates prediction errors yielded by the model and model fitness metrics. If the results are inadequate, this process can be revised to refine models. If prediction performance needs to be improved, this entire process may be repeated. This process iteratively improves the quality of predictions (for time series prediction) or model adjustment (for time series modeling).

Evaluating the performance of a predictive model can be done in several ways. The most common measures are Mean Square Error (MSE) and Symmetric Mean Absolute Percentage Error (sMAPE). The MSE for n predictions is described by Eq. 2.17. The sMAPE is an accuracy measure based on percentage (or relative) errors. For n predictions, it is described in Eq. 2.18 [139], where the absolute difference between actual x_i and predicted $\hat{x}_i$ values is divided by half the sum of absolute values of the actual x_i and the predicted $\hat{x}_i$ values. The Residual Sum Squared (RSS) is the sum of squared differences between actual x_i and predicted $\hat{x}_i$ values (see Eq. 2.19). Finally, R-Squared (R^2), also known as the coefficient of determination, is the proportion of the variance in the dependent variable that is predictable by a model [70]. It is described in Eq. 2.20.

$$\text{MSE} = \frac{1}{n}\sum_{i=1}^{n}\left(\hat{x}_i - x_i\right)^2, \tag{2.17}$$

$$\text{SMAPE} = \frac{100}{n}\sum_{i=1}^{n}\frac{2|x_i - \hat{x}_i|}{\left(|x_i| + |\hat{x}_i|\right)} \tag{2.18}$$

$$\text{RSS} = \sum_{i=1}^{n}(\hat{x}_i - x_i)^2 \tag{2.19}$$

$$\text{R}^2 = 1 - \frac{\sum_{i=1}^{n}(\hat{x}_i - x_i)^2}{\sum_{i=1}^{n}(\overline{x} - x_i)^2} \tag{2.20}$$

For both MSE and sMAPE, the minimum values are 0, and the maximum is infinite (∞). The lower the values, the better the prediction. Regarding R^2, the minimum value is minus infinite ($-\infty$), and the maximum is 1. The higher the value of R^2, the better the prediction. The R^2 may be more informative than MSE and sMAPE for regression analysis [70].

2.4.1 Statistical Models

Simple Differencing can be used together with the linear Autoregressive Moving Average (ARMA) model, producing one of the most important time series linear prediction models, ARIMA [45], which can model stochastic trends [78]. An ARIMA(p, d, q) model is composed of an Autoregressive (AR) and a Moving Average (MA) modeling process (represented by p and q, respectively) with the application of a preliminary Simple Differencing (I) (represented by the order d) to handle nonstationarity in time series [119]. ARIMA models assume that an observation of a time series, x_t, can be described as a function of its p past values and its q past white noise values [259]. The latter is represented as ω_t, a Gaussian white noise series with mean zero and variance σ_ω^2. Let $\theta(B)$ and $\phi(B)$ be the AR and MA operators, respectively, and B is the backshift operator. The ARIMA model is denoted in Eq. 2.21 [45].

$$\theta(B)(1 - B)^d x_t = \phi(B)\,\omega_t. \tag{2.21}$$

Figure 2.17 shows an ARIMA model used for prediction. The time series is split into a training and test set. The test size equals four. The ARIMA model is adjusted using the training size, leading to an ARIMA(1, 1, 3). The adjusted model is plotted as blue dashed lines, and the real observations are plotted as black dots. The last four observations are plotted as red dashed lines corresponding to the four-step-ahead prediction (Fig. 2.17). During model adjustment, the R^2 equals 0.92; during step-ahead prediction, the R^2 equals −1.05.

Other statistical methods are used to model the time series, such as an inherent trend or seasonality. These models may be combined with transformations such as detrending or differencing to achieve stationarity. Examples of models falling into this category are set

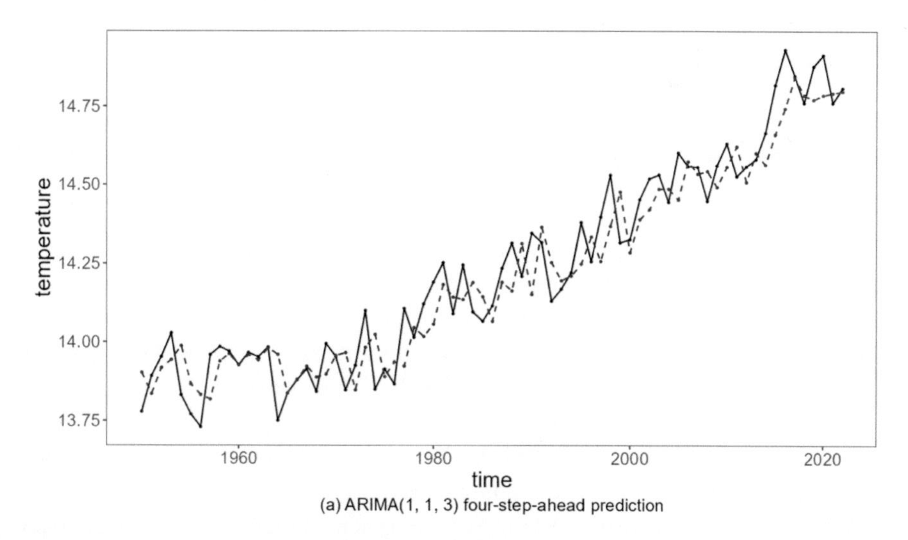

Fig. 2.17 ARIMA prediction model for the training slice of YGT with four-step-ahead prediction

by the Structural Model [73], Holt-Winter's Exponential Smoothing [91], Taylor's Double Seasonal Holt-Winter, Exponential Smoothing State Space Model (ETS), Theta Forecasting, and the Generalized Autoregressive Conditional Heteroscedasticity Model (GARCH).

First, the Structural Model is representative of this category. Structural Model consists of terms such as trend and seasonal. These terms may provide a straightforward interpretation of a time series. A general Structural Model is given in Eq. 2.22, where η_t is the trend term, χ_t is the cycling term, and ω_t is the error term. Trend terms may be stochastic and can be estimated using an ARIMA process.

$$x_t = \eta_t + \chi_t + \omega_t \tag{2.22}$$

Structural Model models may also be considered special cases of state space models and can be represented as such [73]. Thus, their parameters can be estimated by a Kalman filter [308]. A basic example of a state space model is given in Eq. 2.23, which consists of two parts: a so-called state equation (Eq. 2.23a) and an observation equation (Eq. 2.23b). The latter is added since $\dot{x}_t$ (considered a state vector) is assumed not directly observable. Instead, one can only observe x_t, a linear transformation of $\dot{x}_t$ with the addition of noise. Here A_t is a measurement matrix, ϵ is a coefficient matrix, and ω_t and υ_t represent noise [259].

$$\dot{x}_t = \epsilon \dot{x}_{t-1} + \omega_t, \tag{2.23a}$$

$$x_t = A_t \dot{x}_t + \upsilon_t \tag{2.23b}$$

There are several forecasting frameworks based on the classical model of exponential smoothing. A general exponential smoothing model is expressed as in Eq. 2.24, where x_t, obtained by the function of time f, is based on the fitting function vector f_t, the coefficient vector κ_t (T is the transpose operator), and a white noise ω_t. The f_t depends on its past values and the transition matrix M [194].

$$x_t = {\kappa_t}^T f_t + \omega_t, \quad f_t = M f_{t-1} \tag{2.24}$$

Methods based on exponential smoothing, such as Holt-Winter's Exponential Smoothing method, can be used for nonlinear modeling of heteroscedastic time series [91]. The Holt-Winter's Exponential Smoothing method estimates level, slope, and seasonal terms at each time [290]. Despite taking a long processing time to determine a few parameters, it can represent trend, seasonality, and randomness effectively [14].

Another model for estimating heteroscedastic time series is the GARCH. A time series may be explained as a GARCH(p, q) model by Eq. 2.25a, where μ_t is a mean function, σ_t is the conditional standard deviation, and ω_t is a noise series. The ω_t is considered an Independent And Identically Distributed (I.I.D.) variable that follows a normal distribution N(0, 1) [55]. The conditional variance σ_t^2 is defined by Eq. 2.25b with α and β as coefficients [78]. Although GARCH models are useful for treating non-constant variability in time series, they are not able to capture long memory properties and highly irregular behavior [267].

$$x_t = \mu_t + \sigma_t \omega_t \tag{2.25a}$$

$$\sigma_t^2 = \alpha_0 + \sum_{j=1}^{p} \alpha_j \sigma_{t-j}^2 + \sum_{j=1}^{q} \beta_j {x^2}_{t-j} \tag{2.25b}$$

2.4.2 Machine Learning Models

ML have been used for nonlinear time series prediction in many fields [69]. The models generated by ML are universal approximators, as they can approximate any continuous function to arbitrary precision and can be used for modeling time series properties.

Figure 2.18 depicts the general architecture adopted to build ML models, taking a sliding window as input for a set of lagged observations. The number of lagged values is a hyperparameter for ML models, each having a specific set of hyperparameters. The output is the next observation for that sliding window. Thus, it uses the present time series value and its $sw - 1$ predecessors $< x_{i-sw+1}, \ldots, x_{i-1}, x_i >$, to approximate the next value x_{i+1}. The model is built using the training set. If hyperparameters need to be adjusted, the candidate ML model is built by splitting the training set into a novel training and validation set. Once the hyperparameters are established, ML model can be built using the entire training set. The built model can be used later for testing.

The most relevant ML methods are NNET [130], Multilayer Perceptron (MLP) [186], ELM [137], SVM [69], Random Forest Regression [49], and the deep learning models Conv1D and LSTM [128, 172].

A NNET is a bioinspired computational method for recognizing structural data patterns through neurons connected through synapses. The synapses have associated weights representing the relevance of the connection [130]. Usually, NNET has a feed-forward architecture [130]. During a training process, approximation errors are backpropagated to adjust

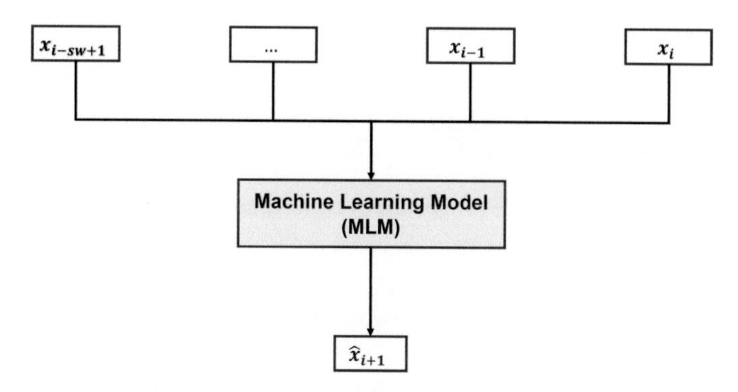

Fig. 2.18 Typical ML model for time series prediction (adapted from Ogasawara et al. [210])

synaptic weights. A practical time series prediction process would involve setting neural network parameters, such as the number of input entries, hidden layers, and neurons in hidden layers [210]. NNET with error backpropagation has been employed for nonlinear time series prediction, outperforming traditional statistical methods such as ARIMA in functional approximation. A fully connected NNET is referenced as MLP, and it is probably the most common network architecture currently in use [186].

The slow learning speed of networks such as NNET and MLP has been a bottleneck in their applications, which is due to slow gradient-based learning algorithms and iterative tuning of network parameters [137]. Unlike traditional implementations, the ELM network adopts a learning algorithm that randomly chooses hidden neuron nodes and analytically determines output weights. Thus, there is no need for any iterative tuning or setting of parameters like learning rate, momentum, or epochs, making learning time very fast [137].

The SVM can recognize patterns in both linear and nonlinear data. Usually, in a regression supported by SVM, a linear learning machine approximates a nonlinear function in a kernel-induced feature space. The system's capacity is controlled by a parameter that does not depend on the dimensionality of the space [130].

Random Forest Regression is based on the combination of decision tree classifiers, or the ensemble of base models, that acts as a "forest". After the formation of the forest, the model may combine the predictions of each tree additively or by average. The results are returned as the estimated time series prediction values [49]. The generalization error for a forest converges if there is a sufficiently large number of trees, which decreases the chance of overfitting.

The idea behind deep learning is to discover multiple levels of representation, expecting high-level resources may represent a more abstract semantics of the data [120]. A Conv1D is an architecture composed of three distinct layers: an input layer, a convolutional layer, and a pooling layer, which reduces the size of the input data. The convolution layers in a time series can be applied as an extractor of characteristics implicit in the data. Equation 2.26 shows a convolution process, where g is the input layer, h is one of the k filters that a Conv1D optimizes for an aim function during the learning process at time t, $*$ is the convolution operator, and n is a hidden layer in the neural network.

$$(g * h)[n] = \sum_{t=0}^{k} g[k-t]h[t] \tag{2.26}$$

LSTM networks have the same properties as conventional recurring networks. However, they can store information for long periods when processing a time sequence. The memory points of an LSTM network are called cells. Cells can carry information until the end of a sequence or identify information the network should forget after some processing step [113].

Figure 2.19 shows a LSTM with four input neurons model for the YGT. Like in the example of Fig. 2.17, the time series is split into a training and test set (size equals four).

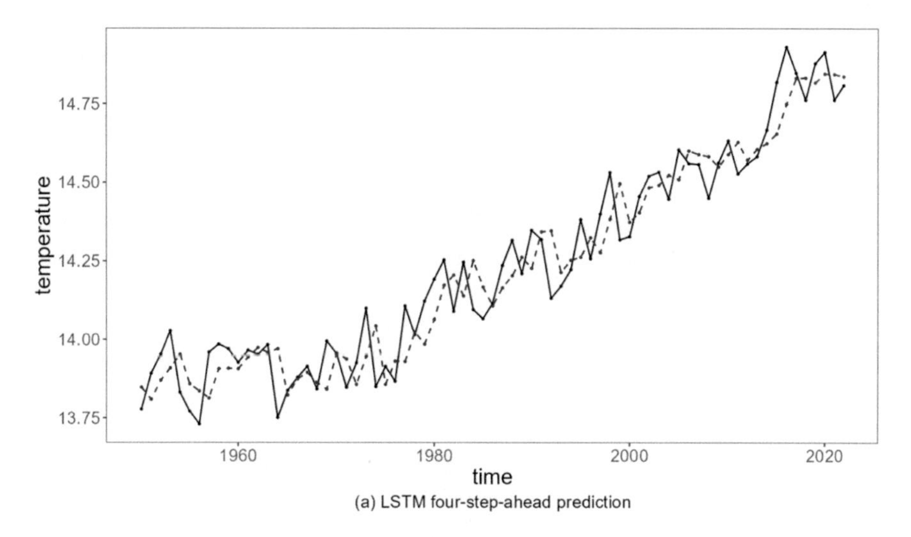

Fig. 2.19 LSTM with four input neurons for YGT with four-step-ahead prediction

The adjusted model is plotted as blue dashed lines, and the real observations are plotted as black dots. The last four observations are plotted as red dashed lines corresponding to the four-step-ahead prediction. During model adjustment, the R^2 equals 0.92; during step-ahead prediction, the R^2 equals -0.13.

It is important to compare ML models with baseline statistical models such as ARIMA to infer the adequacy of the prediction model. Furthermore, ML is also affected by time series preprocessing activities. Thus, during such comparison, given an input time series, values predicted by a particular setup of preprocessing methods and ML models (and their parameters) are compared to the ones found by statistical models. Such benchmarking raises the relative quality of predictions and infers adequate preprocessing-model setups for a particular time series [250]. From examples of Figs. 2.17 and 2.19, ARIMA was comparable with LSTM during training, with R^2 equals 0.92 for model adjustment, but LSTM significantly outperformed ARIMA during test (-0.13 vs. -1.05).

2.4.3 LLMs

The success of Large Language Models (LLMs) in Natural Language Processing (NLP) has inspired adaptations for the time series domain. Using LLMs for time series prediction involves several key aspects. Language and time series models aim to predict future patterns by understanding the sequential structure of the data. In language models, this involves predicting the next token in a sequence, while in time series models, it involves predicting the next values in the series [147].

Time series values are mapped to tokens through scaling and discretization, which, in the context of LLMs, are called quantization. It allows the continuous time series to be represented in a form compatible with language models, which operate on discrete tokens from a fixed vocabulary. Scaling involves normalizing the values in the time series to a common scale to facilitate optimization. Mean scaling is often used, where each value is normalized by the mean of the absolute values in the historical context. Then, quantization discretizes the scaled values into bins, each represented by a unique token, converting the continuous data into a sequence of discrete tokens [30].

Pretrained language models, particularly those based on the transformer architecture, are used without modification. The tokenized time series is fed into these models, which predict the next token in the sequence. The models are trained using the cross-entropy loss, which measures the difference between the predicted token distribution and the actual token. The goal is to minimize this loss, improving the model's ability to predict future values in the time series [128].

The output of the language model is a probabilistic distribution over the possible next tokens. Multiple tokens are sampled from this distribution during inference to generate a probabilistic forecast. These tokens are then mapped back to numerical values through dequantization and unscaled to obtain the forecasted values. LLMs can leverage their pre-training on diverse time series to perform well on new, unseen time series without additional task-specific training, an ability known as zero-shot learning [115].

Synthetic data is generated using techniques like Gaussian processes and data augmentation methods like TSMixup to address the scarcity of high-quality time series. Such data enhances the diversity and robustness of the training data, improving the model's generalization capabilities [30]. These principles allow LLMs to handle the task of time series forecasting by adapting their capabilities in sequence modeling to predict future values in time series [10].

2.4.4 Graph-Based Models

Graph-Based Models for time series are innovative methods that leverage relational information embedded in graphs to support time series prediction. Graph structures represent dependencies among time series, where nodes correspond to individual time series and edges represent pairwise relationships. This method embeds relational information as an inductive bias in the forecasting model, allowing the model to focus on relevant dependencies and local correlations [189].

Spatiotemporal Graph Neural Networks (STGNNs) use the message-passing (MP) framework, where node representations are updated based on information from their neighbors. It facilitates the propagation of temporal and spatial information. Architectures integrate temporal and spatial processing in various ways, such as time-then-space (TTS), space-then-time (STT), and time-and-space (T&S) models [65].

Global models train on collections of time series, sharing parameters across all series to improve scalability and leverage larger time series for better generalization [303]. Local models introduce node-specific components or embeddings to account for local effects and patterns in individual time series. Hybrid methods combine global and local models, balancing the benefits of parameter sharing and node-specific customization [65].

Learnable embeddings capture node-specific information, conditioning the representations on the individual characteristics of the time series. These embeddings are incorporated into the model's encoding and decoding steps to enhance the specificity and accuracy of predictions [180].

Scalability considerations focus on designing architectures that can handle large collections of time series and graphs with many nodes and edges, ensuring computational efficiency. Applications of graph-based models extend to various real-world domains, such as traffic forecasting, energy analytics, and air quality monitoring. These models demonstrate practical utility and effectiveness in diverse contexts, instilling confidence in their potential [324].

2.5 Conclusion

This chapter provided a comprehensive overview of time series analysis, beginning with the fundamental components and progressing through preprocessing techniques and prediction methods. Understanding the key components of time series—trend, seasonality, and residual—is important for event detection and leads to more accurate data analysis. Trends indicate the long-term direction of the series, seasonality captures periodic fluctuations, and residuals account for random variations. Recognizing and decomposing these elements is a key step in accurate data analysis.

The chapter explained the concept of stationarity, which is central to time series analysis. Many methods assume that the statistical properties of the series remain constant over time. A stationary series has a constant mean, variance, and autocovariance. However, real-world time series often exhibit nonstationary behavior due to trends, level shifts, heteroscedasticity, or unit roots. Identifying and addressing nonstationarity is essential to avoid misleading inferences and improve model accuracy.

Data preprocessing is a critical step in preparing time series for analysis. This chapter covered various preprocessing techniques, including data cleaning, temporal aggregation, trend extraction, variance stabilization, detrending, differencing, and decomposition. Each method improves the quality of the data, making it suitable for modeling and analysis. Techniques such as sliding windows and data normalization help manage the temporal structure of the data and ensure consistency across different scales and models. These last two are enabling techniques to support ML methods.

Time series prediction involves using historical data to forecast future values. The chapter outlined a process for time series prediction, from data acquisition and preprocessing to model training, prediction, postprocessing, and evaluation. It presents metrics such as MSE, sMAPE, and R^2, which assess the accuracy and effectiveness of the models.

The chapter also presented statistical and ML models for time series prediction. Statistical models, such as ARIMA, state space models, exponential smoothing, and GARCH, provide frameworks for understanding and predicting time series. ML models, including neural networks, support vector machines, random forests, and deep learning architectures like Conv1D and LSTM, offer tools for capturing nonlinear patterns in the data. In addition to traditional methods, the chapter introduced cutting-edge methods such as LLMs and Graph-Based Models. All elements presented here are building blocks to support the time series event detection.

Anomaly Detection 3

3.1 Anomalies

Anomaly detection has been drawing attention for many years due to its impact in many domains, such as fault diagnosis in industry, water quality, outbreaks in health, and fraud detection [59, 182]. Anomalies are observations that do not fit the typical time series observations and seem to be generated by a different process. However, the actual definition and characterization of anomalies might differ among various works. They are also known as outliers and discords [56, 177].

There is a slight difference between anomaly detection and outlier analysis. In outlier analysis, outliers are treated as undesired data, associated with noise and error [5], and are removed before the main data mining function during data preprocessing [125]. Conversely, anomalies are atypical observations worth studying. Examples include fraud and fault detection, where the goal is to detect and analyze the anomalies [39].

We have built a taxonomy of anomalies, which can be organized according to their type, circumstances in which they occur, their dimension, and interpretation. It is depicted in Fig. 3.1. Furthermore, anomaly detection is driven by how detection models learn from data.

3.1.1 Point and Sequence Anomalies

Anomalies can be classified by their type: point or sequence [40]. A point anomaly is a single observation that significantly differs from the remainder of the time series. Figure 3.2a shows a point anomaly (either in green or red) of the temperature in Celsius for each week in Seattle during 2019. The observation in green corresponds to a labeled anomaly identified by specialists and was correctly detected by Forward And Backward Inertial Anomaly Detector (FBIAD) using a sliding window size of 12 [173]. The observations marked in red correspond to false positives.

E. Ogasawara et al., *Event Detection in Time Series*, Synthesis Lectures on Data Management, https://doi.org/10.1007/978-3-031-75941-3_3

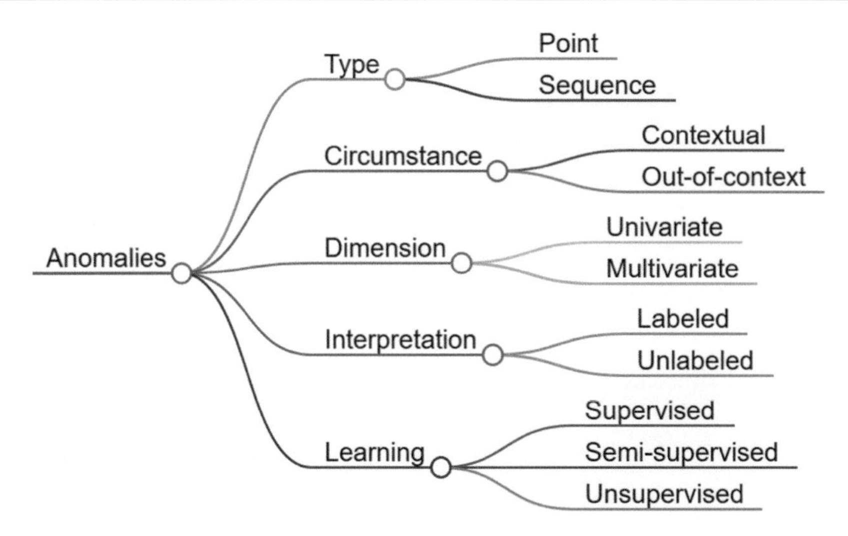

Fig. 3.1 Anomaly taxonomy organized according to type, circumstance, time series dimension, interpretation, and model learning

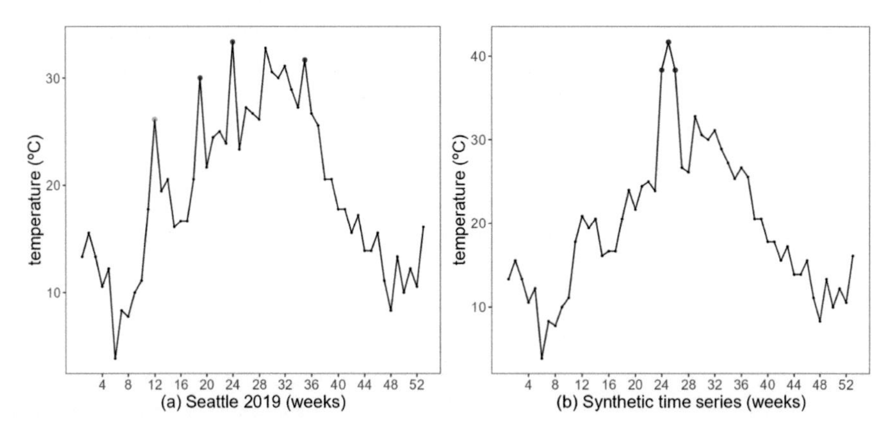

Fig. 3.2 Point and sequence anomalies examples: maximum temperature of each week in Seattle in 2019 (**a**), synthetic time series (**b**)

A sequence anomaly corresponds to a sequence of observations significantly different from the remainder of the time series [39]. Figure 3.2b shows an example of a sequence anomaly of size three (marked in blue) for a synthetic time series. It represents the same temperature data for Seattle in 2019, but we modified the temperatures between weeks 24 and 26. These changes lead to a sequence anomaly from weeks 24 to 26, with atypical high temperatures.

3.1.2 Contextual and Out-of-Context Anomalies

Anomalies can be classified according to the circumstances in which they occur. Consider an anomaly that occurs at time t. It is called a contextual anomaly when the observation at time t is not atypical by itself [203] but is considered anomalous when looking at nearby observations. For example, 26 °C in a day is a common temperature in Seattle during summer. However, such a temperature is anomalous in winter, as reported by local newspapers.[1] An example of a contextual anomaly is shown in Fig. 3.2a, based on data extracted from Weather Underground.[2] Conversely, an out-of-context anomaly is not related to the temporal occurrence of the observation [60], as shown by the blue observations in Fig. 3.2b.

3.1.3 Univariate and Multivariate Anomaly Detection

Anomalies can be classified according to the dimension in which they occur, within univariate or multivariate time series [320]. In univariate time series, an anomaly is an atypical observation that occurs at time t. On the other hand, an anomaly in a multivariate time series can be an unusual combination of observations at a given time t. Figure 3.3a and b exemplifies a multivariate time series scenario, where the shown two variables are generated using the standard normal distribution for illustration. Both series can be analyzed using univariate methods for anomaly detection. The observations in Fig. 3.3a and b marked as red or green correspond to detections obtained using the ARIMA model.

The extreme values in Fig. 3.3a and b are usually anomalies. However, certain combinations of observations can be statistically anomalous but are not captured by individual time series analysis. This scenario opens room for multivariate anomaly detection methods. Figure 3.3c shows a PCA-based anomaly detection method applied to the multivariate time series. The PCA for these two variables is computed, and the residual between observations and the PCA is checked for atypical values, marked in green and red. The blue points in Fig. 3.3a and b correspond to observations that were atypical in the combination of the two series, where observations in green in Fig. 3.3 matched those detected by both univariate and multivariate methods. The observations marked as blue in Fig. 3.3a and b were detected by the multivariate method but not by the univariate method. Conversely, observations marked as blue in Fig. 3.3c were detected by univariate methods but not by multivariate methods. Thus, combining univariate methods is not enough to capture certain anomalies obtained in multivariate analysis. However, multivariate time series anomaly detection does not replace univariate anomaly detection, which should be combined.

---

[1] https://github.com/eogasawara/TSED/wiki.

[2] https://www.wunderground.com.

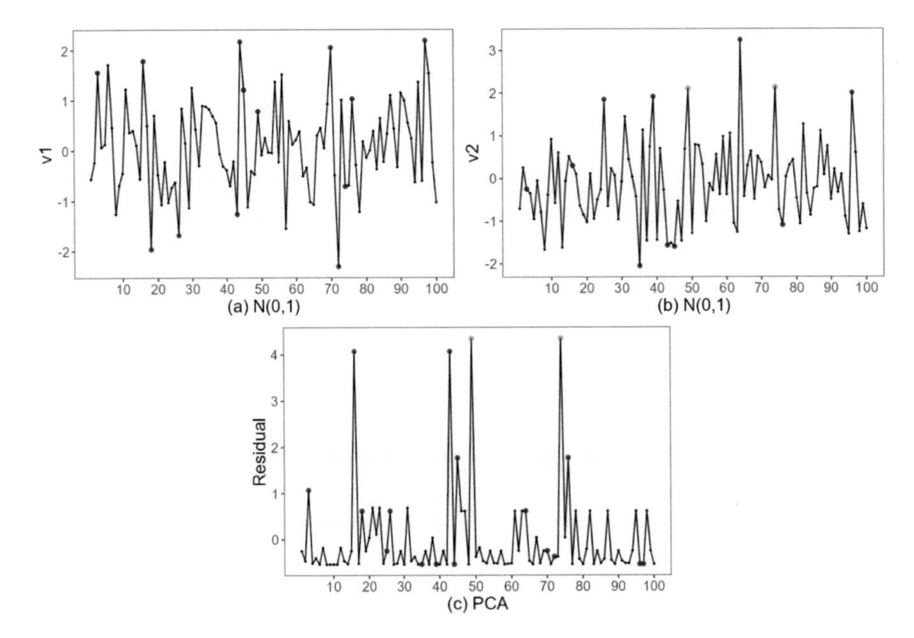

Fig. 3.3 Multivariate example: two standard normal distribution variables (**a**) and (**b**), residual of PCA multivariate anomaly detection method (**c**)

3.1.4 Labeled and Unlabeled Time Series

A time series might have labeled information about each observation, indicating whether it is an anomaly, which is called a labeled time series. However, when there is no such label, it is an unlabeled time series, in which anomalies can be evaluated using statistical analysis, yet subject to interpretation [143]. When the time series is labeled, such labels are commonly provided by human experts. However, there are two difficulties in providing these labels. First, some anomalies might be missed during labeling. Second, the indication of the anomaly might be registered imprecisely (at an incorrect time), leading to a lag between the anomaly and the actual time it occurred [296].

3.1.5 Supervised, Semi-supervised, and Unsupervised

Detection models aim at discovering anomalies and can be classified according to how they learn from data: (i) supervised, (ii) semi-supervised, and (iii) unsupervised [60]. In supervised learning, the time series has a label associated with each observation, indicating if it is an anomaly or typical data. When a model is trained in supervised learning, it is expected to determine the class label of a novel observation. Thus, supervised learning separates the time series into training and testing sets, which is a common practice in data mining [125]. This separation makes sense in online anomaly detection, where the entire time series is used for training and new upcoming data is tested.

There are major challenges in supervised learning for anomaly detection. First, anomalies are, by definition, rare in the time series. This scarcity raises the traditional imbalanced class distribution problem [125]. Techniques such as data augmentation inject artificial anomalies into a time series to obtain a labeled training time series [320] as a way to address this issue. The second challenge concerns obtaining time series with labels, which raises the issues previously mentioned [56].

In semi-supervised learning, the training data is assumed to have only instances of typical data. Since it does not require anomaly class labels, the methods based on semi-supervised learning are more used than methods based on supervised learning [270]. Finally, no training data is required in unsupervised learning, i.e., the methods based on unsupervised learning assume that anomalies are distant from typical observations. Since it is less constrained learning, it is the most adopted one [331].

3.2 Methods for Anomaly Detection

This section covers the main anomaly detection methods, which are grouped into six categories: regression, classification, clustering, statistical, spectral, and information theory. Figure 3.4 describes a synthetic time series that compares methods under a controlled situation and separates the time series into training (between 1 and 75) and testing (between 76 and 101). We have made representative anomaly detection methods available through our publicly Harbinger R package (see Appendix A).

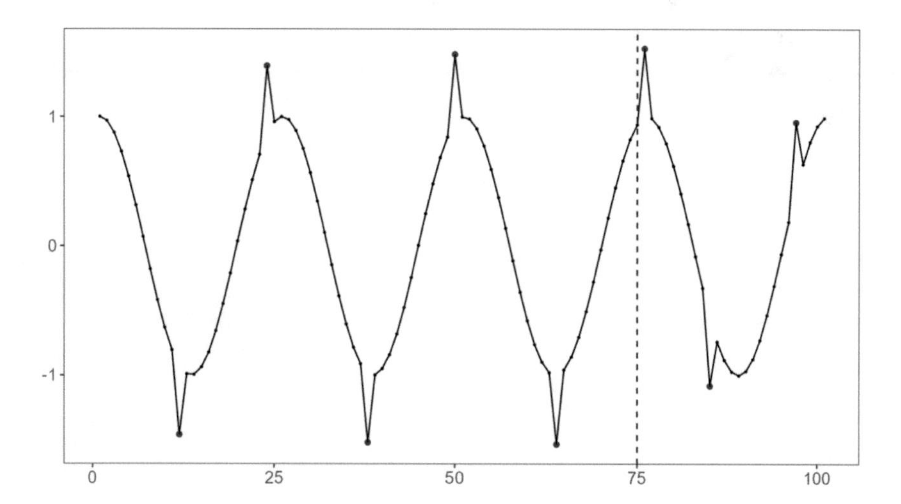

Fig. 3.4 Synthetic time series for comparison of methods (Example 18 available at Harbinger [211])

3.2.1 Regression-Based

Regression-based anomaly detection is based on model deviation, which refers to identifying anomalies where there is a deviation between the actual observations of the time series and the predicted observations by a model. First, a statistical or ML model is fitted to the available data. Then, anomalies are identified as the observations that significantly deviate from the fitted model.

Consider a time series X whose observations are represented as x_t $(1 \leq t \leq |X|)$. A regression model can be built, providing a prediction (or estimation) for each observation $\hat{x}_t$. The distance between observations and estimations can be modeled as a residual time series ω_t, such that $\omega_t = distance(x_t, \hat{x}_t)$. An anomaly is an observation that significantly deviates from the estimation, i.e., $\omega_t > \gamma$, where γ stands for a threshold to model deviations. Regression-based anomaly detection is sensitive to the method adopted to model the time series and the function used to measure the distance between the predicted and observed values [163].

Figure 3.5 shows an example of a time series organized as a sliding window of size 5. The four lagged terms of the time series x_{t-4} to x_{t-1} are used to predict x_t. The prediction is expressed as $\hat{x}_t$. In this case, ω_t corresponds to the difference between $\hat{x}_t$ and x_t. The distribution of ω_t is analyzed for noise anomalies (see Chap. 1).

As expected, the way distance is computed interferes with whether the anomalies are detected. A typical method is to compute the absolute difference between observations and estimations, where the error is relatively easy to interpret and has the same scale as the data. The squared difference between observations and estimations is also well-known and adopted, and since the differences are squared, large errors are highlighted.

Observations not conforming to distance distribution are registered as anomalies in regression methods. The simplest way to identify if an observation is an anomaly is by assuming a Gaussian distribution for the distance [288]. When using the Gaussian distribution, a parametric analysis is assumed, and observations that are distant from the mean $(\overline{\omega} \pm 3\sigma_{\omega})$

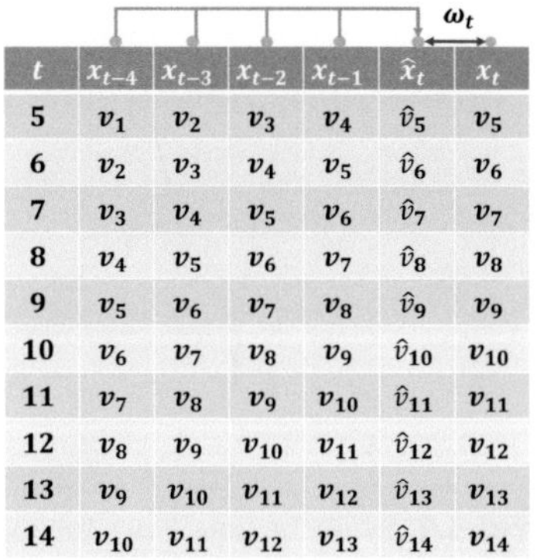

Fig. 3.5 Example of model deviation detection using sliding windows

t	x_{t-4}	x_{t-3}	x_{t-2}	x_{t-1}	$\hat{x}_t$	x_t
5	v_1	v_2	v_3	v_4	$\hat{v}_5$	v_5
6	v_2	v_3	v_4	v_5	$\hat{v}_6$	v_6
7	v_3	v_4	v_5	v_6	$\hat{v}_7$	v_7
8	v_4	v_5	v_6	v_7	$\hat{v}_8$	v_8
9	v_5	v_6	v_7	v_8	$\hat{v}_9$	v_9
10	v_6	v_7	v_8	v_9	$\hat{v}_{10}$	v_{10}
11	v_7	v_8	v_9	v_{10}	$\hat{v}_{11}$	v_{11}
12	v_8	v_9	v_{10}	v_{11}	$\hat{v}_{12}$	v_{12}
13	v_9	v_{10}	v_{11}	v_{12}	$\hat{v}_{13}$	v_{13}
14	v_{10}	v_{11}	v_{12}	v_{13}	$\hat{v}_{14}$	v_{14}

ω_t

(a)

are anomalies, where $\overline{\omega}$ and σ_{ω} stand for the mean and the standard deviation of distance, respectively.

An alternative to the parametric analysis is to use Grubb's test [114], known as the maximum normed residual test, which is a statistical test used to detect anomalies in a time series by identifying data points that are significantly different. It compares the suspected outlier to the sample mean and standard deviation. If the difference between the suspected outlier and the mean is significantly larger than expected based on the standard deviation, the data point is considered an outlier [277].

The box plot rule is the simplest nonparametric statistical test to support detecting anomalies, considering that distant observations are relative to the first quantile (Q1) and third quantile (Q3). The Inter Quartile Range (IQR) equals Q3 – Q1. Atypical observations are not between the interval defined by Q1 – 1.5×IQR and Q3 + 1.5×IQR. This interval contains 99.3% of observations [125]. The lower constraint does not make sense in anomaly detection for regression-based methods, as it stands for very precise models concerning the observations. Thus, only distances greater than Q3 + 1.5×IQR are registered as anomalies.

Extreme value theory is a more advanced method for evaluating distance distribution from the model. It computes a typical time series boundary for observations, as it adopts a density-based comparison to detect any significant changes in the distribution of the observations [273]. This method is adopted in online anomaly detection since it does not make assumptions on the distance distribution [260].

The quality of the model used is also relevant for regression-based anomaly detection. The simplest methods used to model the series include trend models (or followers), which include FFT, Wavelet Transform, Moving Average Smoother, FBIAD [173], Empirical Mode Decomposition (EMD), Refined Empirical Mode Decomposition (REMD) [266], and splines. The first six methods are rigid because their modeling assumption provides inertia [119] while the last one is flexible, adjusting smoothly to the data [145].

Various methods exist to model time series for statistical and ML [239]. The former includes Linear Regression and ARIMA. The latter includes Random Forest Regression, XGBoosting, SVM, MLP, ELM, K-Nearest Neighbors (KNN), Conv1D, and LSTM. These representative methods are covered in Chap. 2. Figure 3.6 exemplifies a regression-based anomaly detector using ARIMA, where the model adjusted to the time series is represented in dashed blue. The residuals at anomalies are significantly different from the remainder of the residuals. Surrounding anomalies, the ARIMA model is slightly influenced, and the distance between the adjusted model and the time series is higher than usual. As the anomaly's influence decreases, the model adjusts to the time series.

Recent works in statistical methods include Triple Exponential Smoothing and Median Method [253]. Triple Exponential Smoothing handles trends and seasonal components by smoothing the data at three levels, whereas Median Method uses the median value of the context sliding window to forecast the next data point. Conversely, recent works in Deep Learning methods include Deep Anomaly Detection (DeepAnT), Deep Network Anomaly Prediction (DeepNAP) [161], LSTM Anomaly Detection, and Telemanom. DeepAnT employs

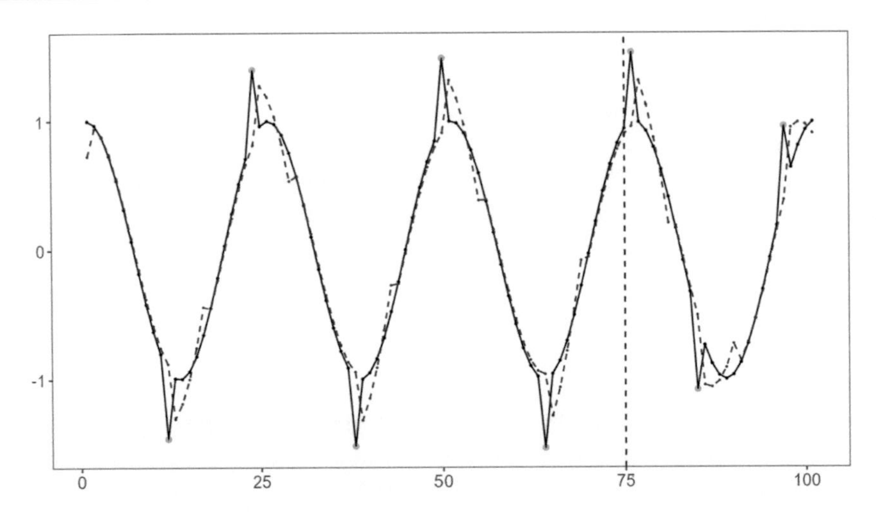

Fig. 3.6 Anomaly detection using ARIMA

Convolutional Neural Network (CNN), whereas LSTM Anomaly Detection [187] and Telemanom [138] are LSTM-based detectors. Health Echo State Networks applies echo state networks [67].

Finally, hybrid and ensemble methods combine multiple methods to leverage each other's strengths for anomaly detection in time series. For example, Hybrid KNN improves upon the standard KNN by combining the strengths of multiple KNN algorithms, leveraging both local and global properties to enhance anomaly detection [263]. Torsk combines statistical and deep learning methods [131]. Similarly, Anomaly Detection-Linear Time Invariant uses multiple linear models to forecast future points, capturing underlying trends and seasonality [300].

3.2.2 Classification-Based

Classification-based anomaly detection involves training a classification model on a labeled time series. Methods in this category perform binary classification of time series observations as falling into a typical or anomaly class based on previously learned patterns. Figure 3.7a shows an example of a time series organized as a sliding window of size 5. During the model building, the labels for the presence of anomalies should be known and used during training. All lagged terms of the time series x_{t-4} to x_t are used to predict the class label for the anomaly e_t. The model's prediction is $\hat{e}_t$. Some metrics for evaluating anomalies, such as f_1, are used to measure the model's accuracy during training, leading to the confidence of the model prediction during testing. As shown in Fig. 3.7b, no labeled data is provided during testing.

t	x_{t-4}	x_{t-3}	x_{t-2}	x_{t-1}	x_t	$\hat{e}_t$	e_t
5	v_1	v_2	v_3	v_4	v_5	$\hat{b}_5$	b_5
6	v_2	v_3	v_4	v_5	v_6	$\hat{b}_6$	b_6
7	v_3	v_4	v_5	v_6	v_7	$\hat{b}_7$	b_7
8	v_4	v_5	v_6	v_7	v_8	$\hat{b}_8$	b_8
9	v_5	v_6	v_7	v_8	v_9	$\hat{b}_9$	b_9
10	v_6	v_7	v_8	v_9	v_{10}	$\hat{b}_{10}$	b_{10}
11	v_7	v_8	v_9	v_{10}	v_{11}	$\hat{b}_{11}$	b_{11}
12	v_8	v_9	v_{10}	v_{11}	v_{12}	$\hat{b}_{12}$	b_{12}

(a)

t	x_{t-4}	x_{t-3}	x_{t-2}	x_{t-1}	x_t	$\hat{e}_t$
13	v_9	v_{10}	v_{11}	v_{12}	v_{13}	$\hat{b}_{13}$
14	v_{10}	v_{11}	v_{12}	v_{13}	v_{14}	$\hat{b}_{14}$

(b)

Fig. 3.7 Example of classification-based detection using sliding windows: **a** during training, labeled data is available; **b** during testing, no labeled data is available

Anomaly detection using classification models is in the category of supervised learning, where Logistic Regression, Naive Bayes, Decision Tree, Random Forest, SVM, NNET, LSTM, and Conv1D being the main traditional classifiers [60, 125]. As previously mentioned in supervised learning, classification is used more for online anomaly detection since it requires training and testing phases. The training phase uses observations and labels to build the model, whereas the testing phase uses the model with new upcoming observations [276]. Online event detection is covered in Chap. 6.

Figure 3.8 shows anomaly detection using an SVM classifier. The model uses SVM with a radial kernel with $\epsilon = 0$ and $cost = 80$. In this example, during training, all anomalies were confirmed, but during testing, only the first anomaly (observation 80) was detected. The other anomaly was marked as a false negative. In this family, Phase-Space Support Vector Machine (Phase-Space-SVM) is a popular method for anomaly detection in time series, in particular when labeled anomalies are rare [185].

3.2.3 Clustering-Based

Clustering is a data mining function that groups similar data instances into clusters [170]. It is based on unsupervised learning [9]. Clustering-based anomaly detection refers to identifying and grouping anomalies that exhibit similar patterns. In this process, a set of features or attributes is extracted from the data, and clustering algorithms are used to group data according to their similarity. Outlier samples are identified through a clustering method by comparing them against typical sample clusters. The clustering-based anomaly detectors assume anomalies are related to outlier samples [192].

Figure 3.9 shows an example of clustering-based anomaly detection using a sliding window, where each sliding window is assigned to a cluster $\ddot{r}_c$. Each cluster has a representative sequence for distance evaluation, such as a centroid. All sliding windows are evaluated

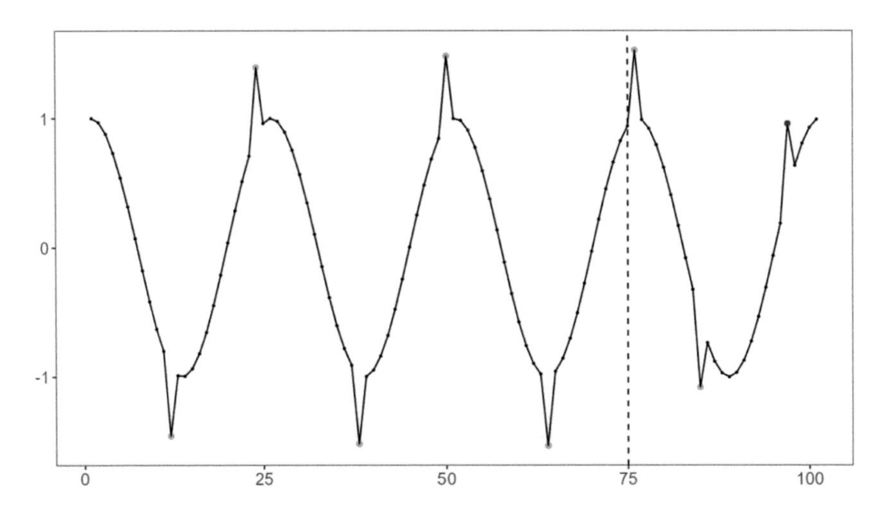

Fig. 3.8 Anomaly detection using SVM classifier

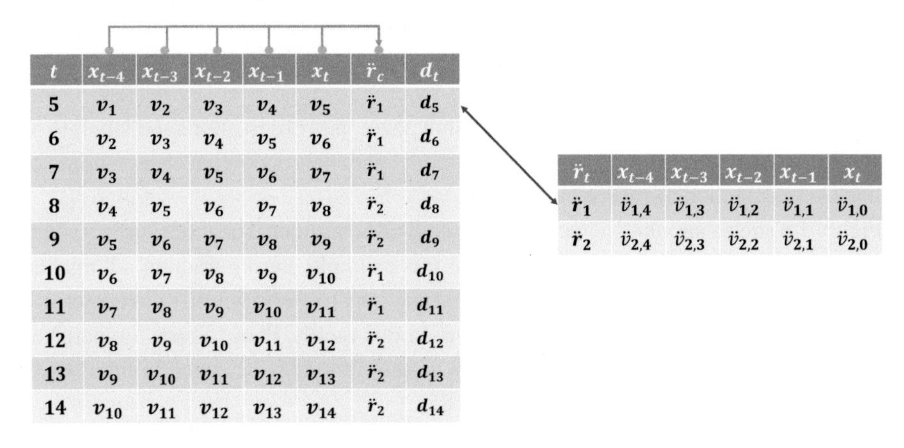

t	x_{t-4}	x_{t-3}	x_{t-2}	x_{t-1}	x_t	$\ddot{r}_c$	d_t
5	v_1	v_2	v_3	v_4	v_5	$\ddot{r}_1$	d_5
6	v_2	v_3	v_4	v_5	v_6	$\ddot{r}_1$	d_6
7	v_3	v_4	v_5	v_6	v_7	$\ddot{r}_1$	d_7
8	v_4	v_5	v_6	v_7	v_8	$\ddot{r}_2$	d_8
9	v_5	v_6	v_7	v_8	v_9	$\ddot{r}_2$	d_9
10	v_6	v_7	v_8	v_9	v_{10}	$\ddot{r}_1$	d_{10}
11	v_7	v_8	v_9	v_{10}	v_{11}	$\ddot{r}_1$	d_{11}
12	v_8	v_9	v_{10}	v_{11}	v_{12}	$\ddot{r}_2$	d_{12}
13	v_9	v_{10}	v_{11}	v_{12}	v_{13}	$\ddot{r}_2$	d_{13}
14	v_{10}	v_{11}	v_{12}	v_{13}	v_{14}	$\ddot{r}_2$	d_{14}

$\ddot{r}_t$	x_{t-4}	x_{t-3}	x_{t-2}	x_{t-1}	x_t
$\ddot{r}_1$	$\ddot{v}_{1,4}$	$\ddot{v}_{1,3}$	$\ddot{v}_{1,2}$	$\ddot{v}_{1,1}$	$\ddot{v}_{1,0}$
$\ddot{r}_2$	$\ddot{v}_{2,4}$	$\ddot{v}_{2,3}$	$\ddot{v}_{2,2}$	$\ddot{v}_{2,1}$	$\ddot{v}_{2,0}$

Fig. 3.9 Example of clustering-based detection using sliding windows

according to a distance metric (such as Euclidean Distance) concerning the representative sequence. The distance distribution analysis characterizes the presence or absence of an anomaly. As it computes the distance of the entire sliding window, Clustering-based anomaly detection can be used for both point and sequence anomaly detection.

Several clustering-based anomaly detection methods have been developed. They can be classified into two major categories: (i) density-based and (ii) distance-based [81, 253]. The density-based category assumes that typical observations belong to a cluster. Conversely, anomalies do not belong to any cluster [81]. This assumption relies on clustering methods that are not obliged to assign a cluster to each observation. These methods are optimized to find similar data. The ones that do not fit such similarities are anomalies. More formally, methods within this category consider that a sequence w (of size l) with less than κ close neighbors are anomalies. Close neighbors are the ones with a distance smaller or equal to γ. In this way, w_t is an outlier in a time series X if and only if $|\{\forall w \in sw(X, l), d(w, w_t) \leq \gamma\}| < \kappa$ [39].

Density-based methods include Density-Based Spatial Clustering of Applications with Noise (DBScan) [96], ROCK [118], DBStream (DBStream), Local Outlier Factor (LOF), and Connectivity-Based Outlier Factor (COF). This category also includes methods that separate typical data from anomalies based on the size of the clusters. Typical data belongs to large clusters, whereas anomalies belong to smaller clusters. The DBStream is tailored for evolving data streams and uses a density-based method to form micro-clusters, which are then merged into larger clusters over time. This method computes anomaly scores based on the distance between data points and their closest micro-cluster centroids, making it suitable for dynamic environments like network traffic monitoring and financial fraud detection due to its ability to adapt to new patterns and discard outdated information [124].

LOF compares the local density of a data point to that of its neighbors, identifying points with significantly lower density as anomalies. This local density comparison makes LOF quite effective in detecting anomalies in time series with varying density distributions [50]. Conversely, COF measures the average chaining distance to capture the connectivity of a point with its neighbors. Points with lower connectivity are considered anomalies, making this method effective in detecting anomalies in time series with irregular structures [34].

The distance-based category assumes that typical data are close to clustering representatives. A common method considers centroids, such as K-Means [144], as the clustering representatives. The centroid of each cluster is defined by the algorithm iteratively by averaging sequences within each cluster [39]. Anomalies are distant from their closest cluster centroid. Thus, the method consists of clustering data and computing the distance between observations and their respective centroids. Medoids are also an alternative for the clustering representatives adoption [152].

Figure 3.10 shows an example of anomaly detecting using clustering. It applies a K-Means using k and sequence size equals three and one, respectively. Since it is an unsupervised learning method, it computes the centroids using the entire time series. Sequences that are far away from centroids are marked as anomalies. It correctly found five out of the eight anomalies presented in the example.

This category also supports semi-supervised learning. In this case, typical data are clustered using Self-Organizing Maps (SOM) and Expectation Maximization (EM) [64, 170]. This category organizes clusters of typical data. Then, new observations are compared against the clusters. An anomaly score is obtained for them. This is why the training time series should not have anomalies. Otherwise, these methods would be unable to detect them [39]. Also, clustering is sensitive to the distance adopted. Other measures, such as the well-known Dynamic Time Warping (DTW) [86], can replace traditional Euclidean Distance. See Chap. 5 for details on DTW.

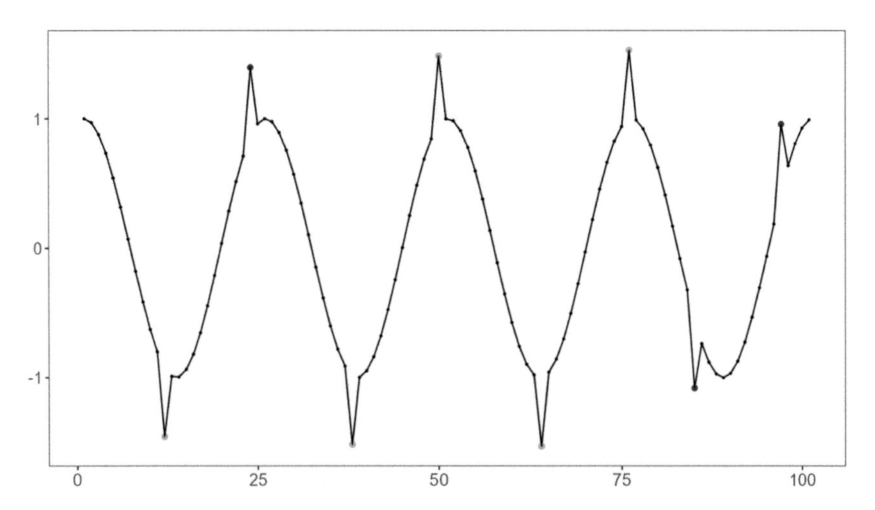

Fig. 3.10 Anomaly detection using K-Means with k equals 3 and sequence size equals 1

There is a broad new set of scalable anomaly detectors for clustering-based detectors, such as Scalable Adaptive Online Nonparametric Anomaly Detection (SAND) and Singular Spectrum Analysis (SSA). They are designed to handle large time series and real-time applications. SAND uses a nonparametric method to model typical observations and dynamically adjusts its parameters based on incoming observations, making it quite effective for real-time applications where data evolves. This adaptability enables SAND to maintain high accuracy in detecting anomalies in dynamic environments [43]. Conversely, SSA decomposes time series into interpretable components such as trend, oscillatory components, and noise. It then reconstructs the series to identify anomalies as deviations from the expected pattern [309].

3.2.4 Statistical-Based

Statistical-based anomaly detection methods allow for identifying anomalies or situations deviating from the expected observations, involving the analysis of the data and using statistical tests to detect changes, anomalies, or patterns indicative of potential problems or failures. They may include time series analysis, hypothesis testing, and regression analysis.

The main principle of statistical anomaly detection is that typical observations occur in high-probability regions of a statistical model while anomalies occur in the low-probability regions [6, 145]. Models can be either parametric or nonparametric. For parametric models, there are two basic steps. In the first step, the statistical model is fitted to the data. There are many measures to evaluate model fitting, including MSE, sMAPE, R^2, and AIC [46, 288].

In the second step, the residual for each test instance is used to determine anomalies, based on the same principles of regression-based models, but using statistical assumptions [48].

Under the nonparametric model analysis, the histogram-based methods fall into the simplest methods, using histograms to maintain a profile of typical observations. This method is also known as frequency-based or counting-based. Like classification models, they are more appropriate for online event detection. The histogram is built using the training set, and new observations are tested to determine whether they fall into any one of the bins of the histogram. The observation is typical if it falls in a frequent bin or is anomalous otherwise [60]. The bin intervals used to build the histogram are key for anomaly detection. If they are too short, many typical observations fall in empty or rare bins, resulting in a high false alarm rate. If they are too large, many anomalous observations might fall in frequent bins, resulting in a high false negative rate [39].

The histogram-based methods have limitations, especially when dealing with high-dimensional or multi-modal data. In such cases, creating histograms becomes challenging as it may require dividing the data into many bins, leading to sparse histograms and loss of detail. These methods might find difficulty in capturing the relationship among dimensions [60].

Figure 3.11 shows anomalies detected using the histogram method. Anomalies are observations that fall outside bin intervals or observations present in bins with low frequency. In the former criteria, anomalies are values less than -2 or greater than 1.5. In the latter, anomalies are values in bins 1, 2, and 7 (between -2 and -1 or 1 and 1.5). Some of the anomalies in the training set were removed from the example of Fig. 3.11 since they would be too frequent. Due to their high frequency, the histogram would fail to find them in such a case.

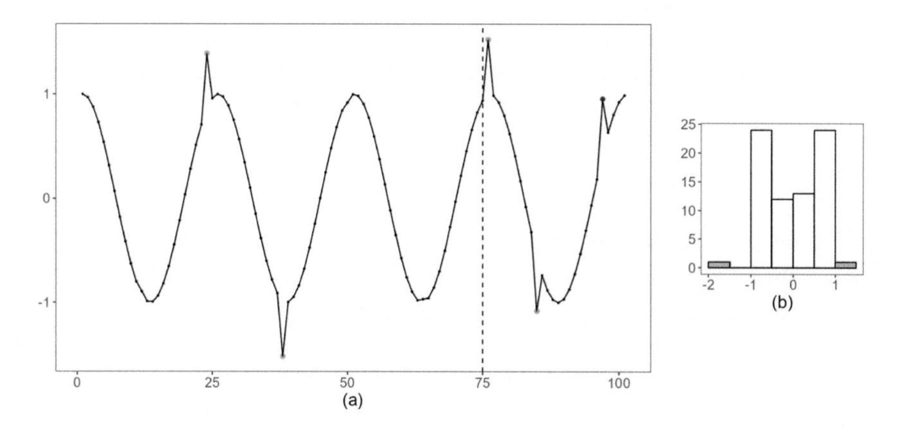

Fig. 3.11 Anomaly detection using histogram (**a**). Computed histogram in the training set to support anomaly detection (**b**)

An advantage of statistical methods is that they adopt statistically justifiable solutions for anomaly detection. An additional benefit is their ability to define a confidence interval that separates anomalies from typical observations. Conversely, the disadvantage of statistical methods is the implicit assumption that the data is generated from a particular distribution, which often does not hold, especially for nonstationary or high-dimensional real-world time series [48]. Advanced statistical-based methods include the ones based on Probability Density Function (PDF). The estimated PDF models the typical observations. A new observation in a low-probability area of the PDF is declared an anomaly [140].

3.2.5 Spectral-Based

Spectral-based anomaly detection methods try to find an approximation of a time series using a combination of attributes that capture the main variability of the data. This category of methods is based on the premise that data can be embedded into a lower-dimensional space. In this reduced subspace, typical observations and anomalies appear significantly different. Thus, spectral anomaly detection methods adopt the general principle of determining subspaces (such as embeddings and projections) where the anomalous instances can be easily identified [92]. Such methods can work in an unsupervised and semi-supervised setting [60].

Intuitively, spectral-based anomaly detection is suitable for analyzing multivariate time series, in which case, the embedding naturally reduces the number of dimensions. However, it is also applicable to univariate time series. Considering the sliding window sw for a time series of size p, each lagged term is expressed as an individual dimension. Thus, lagged terms could be embedded similarly to multivariate time series.

One of the simplest spectral methods is to use PCA to project data in lower-dimensional spaces [2]. The goal is to analyze the projection of each data instance along the principal components, a typical instance satisfying the correlation structure of the data and the projection. On the other hand, anomalies significantly deviate from the PCA representation [39, 220]. An example of anomaly detection using PCA is shown in Fig. 3.3c.

Recently, autoencoders, i.e., neural networks that learn the most significant features of a training set [8], have become the most used spectral methods [192, 219, 234]. Since anomalies often correspond to non-representative features, autoencoders fail to reconstruct them, providing distances between encoded and original sequences [3]. Figure 3.12 shows an example of an autoencoder anomaly detection method with a neural network of five inputs. The output layer produces five observations, which should be the same as the input. The key point of the autoencoder neural network is the hidden layer structure, which has an hourglass shape. This property is the main principle of this spectral-based anomaly detection, as the encoder layer with fewer neurons than the input has enough information to decode the output values. Anomalies are observations output from the autoencoder that are significantly different from the input. Because of that, many methods train the neural network using semi-

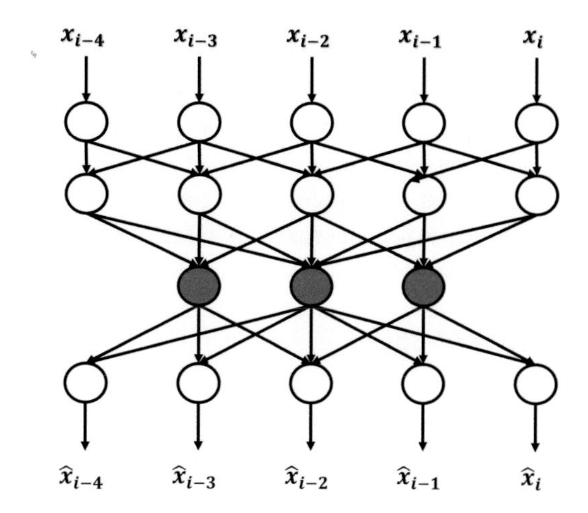

Fig. 3.12 Autoencoder neural network

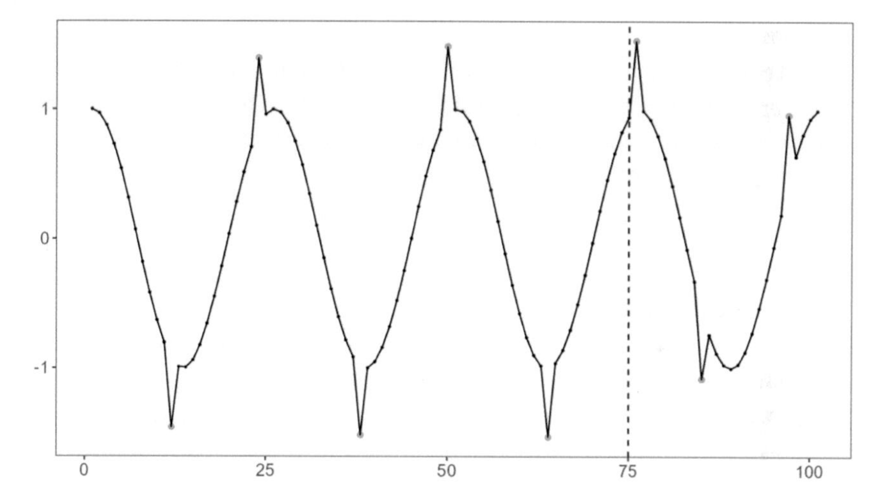

Fig. 3.13 Anomaly detection using an autoencoder with input size equals 3 and encoder layer size equals 1

supervised learning, i.e., the training time series has only typical observations. Each layer might be fully connected to the following. However, for the sake of readability, the example of Fig. 3.12 is not fully connected.

Figure 3.13 shows an example of anomaly detection using an autoencoder. It uses an input size of three neurons, an encoder layer of size 1, and an output size of three. Since it is based on reconstruction error, even though the last two anomalies of the time series were not present in the training set, the model could identify them.

Many new deep learning methods are based on autoencoders, including LSTM-based autoencoders [219, 234] and Variational Autoencoder with a Gated Recurrent Unit. The input of the model is a sequence of observations containing x_t and p preceding observations ($< x_{t-p+1}, \ldots, x_t >$). The output is the reconstructed ($< \hat{x}_{t-p+1}, \ldots, \hat{x}_t >$) [268]. It is also possible to extract features within overlapping sliding windows (e.g., statistical features) before applying the autoencoder to account for temporal dependencies [158].

3.2.6 Information Theory-Based

Information theory-based methods analyze the time series context based on information theory concepts, such as Kolmogorov Complexity and entropy [36, 60]. The general assumption is that anomalies are rare, and typical observations provide little information, whereas anomalies introduce more information and entropy to the sample analyzed.

The general principle can be formalized as follows. Let S be a set of all sequences of size p of a time series X, i.e., $S = sw(X, p)$. The information of these sequences can be represented as $I(S)$. Anomaly detection methods based on information theory aim to find a subset A of S such that $|I(S) - I(S - A)|$ is maximized. The sequences present in A are considered anomalies. To compute the information I, the size of the compressed data file (using any standard compression algorithm) can be used to measure the Kolmogorov Complexity [155].

3.3 Advanced Topics

3.3.1 Volatility Anomalies

Volatility refers to the degree of variation or fluctuation in a time series, which can be measured using statistical methods such as standard deviation or variance. Volatility anomalies are unexpected or atypical fluctuations in the volatility of a time series. Thus, capturing the estimated instantaneous volatile each time becomes important, which can be achieved with models such as Autoregressive Conditional Heteroscedasticity Model (ARCH) and GARCH (see Chap. 2). The latter is an econometric model that is the most well-known and applied model for addressing volatility [55].

Volatility anomalies (va) can be formalized by specializing Eq. 1.2 introduced in Chap. 1 to the volatility component, as described in Eq. 3.1. In this case, an event identified at time t can be considered a volatility anomaly if TC for the volatility component (v) escapes the expected volatility before ($ep(v(x_t))$) and after ($ef(v(x_t))$) instant t.

$$va(X, k, \sigma) = \{t, |v(x_t) - ep(v(x_t), k)| > \sigma \land |v(x_t) - ef(v(x_t), k)| > \sigma\} \quad (3.1)$$

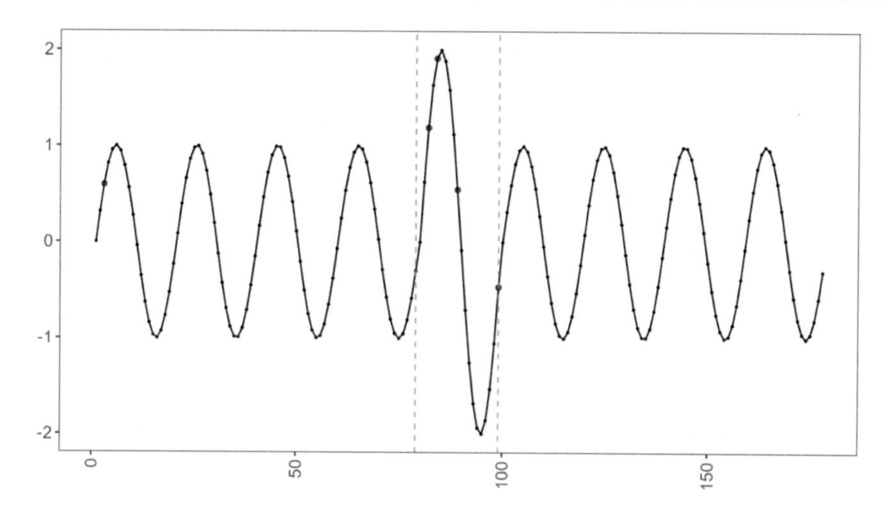

Fig. 3.14 Synthetic time series with increased volatility between observations 79 and 99

Figure 3.14 shows a synthetic time series example with a sequence of observations with increased volatility, with a GARCH anomaly detector. The model finds some anomalous observations within the sequence limited by the two dashed lines, but a false positive is also detected at the beginning of the time series.

Studying volatility anomalies requires methods to identify subtle patterns and relationships that traditional methods might miss. For example, where high volatility is associated with risk in finance, most time series exhibit nonlinear properties, as their volatility varies widely. Therefore, detecting volatility anomalies is an enabling tool to support risk management systems.

3.3.2 Multivariate Time Series

In many cases of multivariate time series, the data might come from different sources or contain different attributes. Multivariate anomaly detection goes beyond univariate methods by considering the relationships and dependencies among multiple variables. Research in methods that handle multivariate data have increased in recent years, because of the importance of considering interdependencies between variables to detect anomalies [99].

A baseline method for multivariate analysis is to use univariate methods to detect anomalies in each time series. The observation at time t is considered typical only if all individual time series do not detect an anomaly at time t. It is considered anomalous if any time series contains an anomaly at time t. Although simple, this baseline method leverages the strengths of univariate analyses while applying them in a multivariate context [268].

Regression-based methods for multivariate anomaly detection model the typical observations [287]. Observations that deviate significantly from the model are indicated as anomalies. Such methods integrate multiple variables, capturing their relationships to detect anomalies more effectively. In this group of methods, KNN has been used for anomalies based on the distance of observations to their nearest neighbors in the dimensional space.

Clustering methods, either distance-based or density-based, have also been used. Anomalies are detected as observations that do not belong to any cluster, the ones that belong to small clusters, and those that are significantly distant from representative observations in each cluster [170].

Spectral-based methods are used to reduce dimensionality and provide feature learning. PCA is adopted to enable univariate time series analysis when the dimensions are reduced to one. Even when the reduced dimensions are still greater than one, both univariate and multivariate methods can be applied [148]. Besides, autoencoders have been adopted to identify anomalies by measuring the reconstruction error concerning the original observations [219]. Autoencoders are able to address complex data that have nonlinear relationships.

Some multivariate time series can be challenging to analyze, requiring more advanced methods. This is the case for spatial-time series, encompassing data that evolve in space and time [99]. In this context, the goal is to identify anomalies that significantly deviate from typical spatial–temporal observations [26, 265]. When observations are related to a moving object, the goal is to discover the most anomalous trajectories [166, 315, 321]. These advanced methods are used in applications such as surveillance and transportation, where understanding movement patterns is essential.

Research papers on this topic often benchmark univariate or multivariate algorithm families in mixed time series (univariate and multivariate), which can obscure whether multivariate solutions are superior to multivariate time series. When a comparison of methods is made using univariate and multivariate time series, such benchmarks fail to highlight the strengths of multivariate anomaly detection algorithms. Works, such as CoMuT, introduce multivariate time series with correlation anomalies and are better suited to showcase the advantages of multivariate methods [297]. Furthermore, the results show that univariate solutions excel at detecting point and sequence anomalies, indicating that effective anomaly detection should employ univariate and multivariate algorithms to detect all anomalies comprehensively.

3.3.3 Graph-Based Methods

Graph Neural Networks (GNN) are methods for analyzing multivariate time series by leveraging the data structure as a graph, capturing the relationships between variables, and treating each variable as a node and the relationships as edges. Multivariate Time series Anomaly Detection using Graph Attention Network (MTAD-GAT) is a representative method that uses this data structure [85, 323]. However, GNN-based methods face challenges when trained on data with anomalies, as they may overfit to noisy patterns if the training data lacks annotations for anomalies [327]. Recent work introduces a technique to filter out anomalies before training [28].

Fused Sparse Autoencoder and Graph Net (FuSAGNet) combine Sparse Autoencoder and Graph Neural Network to model the relationships within multivariate time series [127]. FuSAGNet provides sparse latent patterns from high-dimensional time to improve the graph-based forecasting model. Such combining enables anomaly detection of high multidimensional time series. Graph-Augmented Normalizing Flow (GANF) is another method that integrates GNN with distribution analysis to identify anomalies in multivariate time series [79]. Such a wide variety of ways to model multidimensional time series provides novel opportunities to tackle spatial–temporal and multivariate time series anomaly detection.

3.3.4 Extreme Values

A particular type of anomaly occurs in a time series where some values are considered above or below some thresholds. However, they refer to significant observations of the monitored phenomenon. Observations falling into this category are named after extreme values and are considered to be generated by some extreme event process. Examples of extreme events include extreme meteorological conditions, such as heatwaves, strong rainfall, and tsunamis, leading to extreme wave heights. From a probabilistic point of view, extreme events are naturally scarce and have low frequency. They are denoted by a few point values in the historical curve of the monitored phenomenon. When events are independent of each other or happen at least so far apart to keep any dependency, they are considered I.I.D.. As an example of application, a city hall may plan to build a bikeway at height along its beach coast. To avoid being hit by the waves, estimating the value for the highest waves that could reach the coast would be interesting so that bikeway riders are secured.

Given the low frequency of extreme events, a distribution function that models the underlying phenomenon adopting the representation of a Gaussian distribution places extreme values on both sides of the distribution tails. Thus, estimating extreme values from a history of measurements suffers from the effects of the central limit theorem and tends to be biased by more frequent normal measurements. In this context, dealing with anomalies of extreme values requires a special method. A straightforward method is to extract extreme values from the set of historical observations computed as a max function described in Eq. 3.2, where n is the number of observations in a block, and x_i is an observation, for instance, the daily maximum wave height at day i. Moreover, x_i is univariate, so we can easily define an ordering among its values.

$$M_n = max\{x_1, x_2, \ldots, x_n\} \tag{3.2}$$

The probability of an extreme event to occur, given Eq. 3.2 and a bound on the normal values z, can be computed as in Eq. 3.3 where F(z) is a distribution function, which is, unfortunately, unknown. An alternative is to compute an approximation of $F(z)$ given by the extreme data only.[3]

$$Pr(M_n \leq z) = Pr\{x_1 \leq z\} \times \cdots \times Pr\{x_n \leq x\} = F(z)^n \quad (3.3)$$

Three extreme distributions functions (known as **Gumbel, Fréchet, and Weibull** distributions) can be generalized using the Generalized Extreme Value (GEV) family of distributions [74], depicted in Eq. 3.4. The three variations of Generalized Extreme Value are obtained by considering different shapes for the distribution's tails, given by the ζ shape parameter. While G decays exponentially for the Gumbel distribution, it follows a polynomial shape for the Fréchet one.

$$G(z) = exp\left\{-\left[1+\zeta\left(\frac{z-\mu}{\sigma}\right)\right]^{\frac{-1}{\zeta}}\right\} \quad (3.4)$$

In addition to the ζ shape parameter, the model includes a location μ and a scale parameter σ to fit the three extreme distribution functions. The GEV distribution with its parameters models the distribution of extreme values in time series. The extreme values are extracted from the studied time series using either the *Block* or the *Peak over Threshold* methods. The *Block* method considers a set of m blocks of size n. For each M block, $M_m = \{M_{n,1}, M_{n,2}, \ldots, M_{n,m}\}$, an extreme value is computed, for instance by evaluating a *Max* or *Min* function on the block values. Alternatively, the *Peak over Thresholds* (PM) extraction method selects values above or below a tuned threshold. The extraction process produces a time series of extreme values $Z = \{Z_1, Z_2, \ldots, Z_m\}$.[4]

We exemplify this process by considering a scenario of forecast of extreme rainfall in the city of Rio de Janeiro [231]. We build a dataset comprising the precipitation volume captured by a rain gauge from the city administration, at 15-minute intervals, from 2019 to 2022. Figure 3.15 exemplifies the results of applying a PM extraction method. The threshold was set to a value of 5mm/observation.

Once the extreme value time series have been computed, we can fit the GEV function to the series, optimizing an estimate loss function, such as Maximum likelihood estimate [256].

Finally, the model estimates the probability of the extreme event occurring for a list of return periods, i.e., the probability of the extreme events as the period of consideration increases. In our example of Fig. 3.15, the return value for each return period is depicted in Table 3.1.

The extreme value theory described here exemplifies a particular scenario where anomalous values in time series compose a new time series interpreted as generated by a different and relevant process for decision-makers in various domains such as risk minimization and disaster prevention. In the context of the taxonomy depicted in Fig. 3.1, extreme value time series are classified as univariate and unsupervised.

[3] The model can equally represent minimum values by adapting the aggregating function.

[4] For $\zeta \neq zero$.

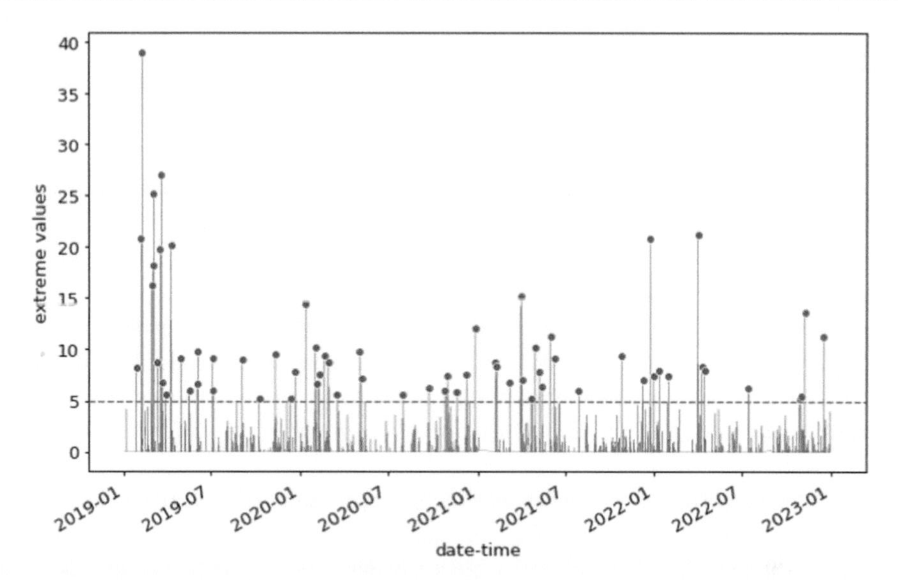

Fig. 3.15 Extreme values extracted from a rainfall dataset covering the period of 2019–2022 from a rain gauge in Rio de Janeiro City. The Extracted values were computed using the *Peak over Threshold* method

Table 3.1 Return values for increasing periods of 30 days. For a period of 2×30 days, the probability of exceedance of having an extreme event of rainfall is 8.437%

Return period	Return value
1	5.470
2	8.437
5	12.366
10	15.336
25	19.262

3.4 Further Readings

Due to the importance of this subject and its numerous applications, several surveys on anomaly detection have been published, especially in the last decade [42, 76, 212, 277, 325]. Some surveys target specific data structures, such as sequences [61] and graphs [13, 240]. On the other hand, others target data characteristics, such as categorical data [270], high-dimensional big data [277], and sequence anomalies [62]. Some works focus on specific methods, such as probabilistic methods [19] and deep learning [18, 81, 214], while others are driven to a particular context, such as finance [11], autonomous electric vehicles [90], power distribution network [328], and outlier analysis from the computer science perspective

[6]. Some works combine the benefits of many different methods using ensembles [52, 167, 238, 329]. Finally, some surveys focus on benchmarking a broad range of anomaly detection methods [4, 11, 121, 143, 202, 215–217, 253, 254].

3.5 Conclusion

This chapter presented the fundamentals of anomaly detection. It distinguished anomaly detection from outlier analysis, noting that anomalies are unusual observations of interest, whereas outliers are often considered noise. The chapter also explained the difference between point anomalies, single observations that deviate significantly from the rest, and sequence anomalies, which involve a sequence of observations that differ from the rest of the time series.

The chapter described contextual anomalies, which are anomalies based on their context, and out-of-context anomalies, which are unrelated to temporal occurrence. It also differentiates between univariate and multivariate anomaly detection, illustrating how multivariate methods can detect anomalies missed by univariate methods. The section on labeled and unlabeled time series explained how labeled time series have information indicating whether each observation is an anomaly, while unlabeled time series lack this information and rely on statistical analysis for anomaly detection.

The chapter covered anomaly detectors based on supervised, semi-supervised, and unsupervised learning methods. Supervised learning uses labeled data to build models that predict anomalies, while semi-supervised learning uses labeled data for typical observations. On the other hand, unsupervised learning assumes typical data is more frequent than anomalies. The methods for anomaly detection were grouped into six general categories: regression, classification, clustering, statistical, spectral, and information theory-based, each with practical applications. Each method was explained with examples to illustrate its application in detecting anomalies.

Regression methods are based on model deviation detection, where significant deviations from predicted observations identify anomalies. Classification methods use ML algorithms to classify observations as typical or anomalous based on learned patterns. Clustering methods group similar data instances and identify anomalies based on their distance to cluster centroids or absence in any cluster. Statistical methods use tests and models to detect anomalies, while spectral methods project data into lower-dimensional spaces to identify anomalies. Information theory-based methods analyze anomalies based on their information content.

The chapter also covered advanced topics in anomaly detection. It introduced methods for detecting volatility anomalies in time series using models like GARCH and described detection methods for multivariate time series. The chapter highlighted the benefits and differences of univariate and multivariate methods. It explored graph-based methods for anomaly detection in multivariate time series, leveraging relationships between variables modeled as a graph. Finally, it focuses on detecting extreme values in time series, presenting methods like the GEV family of distributions to model and estimate these values.

As shown in this chapter, anomaly detection in time series is a vibrant research area with a constant proliferation of novel methods. This dynamic is due to the increasing complexity and volume of data generated in finance, healthcare, IoT, and social networks, which demand more precise methods. However, anomaly detection still presents challenges like scalability, result interpretation, and adaptation to different contexts and data types.

Change Points and Concept Drift Detection 4

4.1 Change Points

Detecting changes in time series, observed as changes in dynamics (distribution or autocorrelation), is important for effective monitoring. When changes are abrupt, the observation that separates each time series interval with different dynamics is considered the change point. However, identifying the exact observation where a change occurs may not be clear or feasible when changes are gradual. Change point detection involves finding these observations that mark the transition between different time series dynamics [23].

Figure 4.1 shows a time series with four change points (cp_{AB}, cp_{BC}, cp_{CD}, cp_{DE}), separating the time series into five segments (from A to E), each corresponding to a different time series dynamic. The cp_{AB} characterizes a change in volatility, the cp_{BC} marks an abrupt change, and the last two change points correspond to gradual changes.

There is a need to distinguish between offline and online change point analysis. In an offline scenario, the goal is to detect when different time series dynamics occur, while in online analysis, the aim is to identify changes in the time series dynamics as quickly as possible. This distinction is important for guiding reactions to undesirable phenomena. Additionally, labels that characterize anomalous observations or change points might not be immediately available in online analysis after the data is produced [271].

Another key aspect that drives most change point detection methods is the prior knowledge of the number of change points in the time series. If the number of change points is known, the task becomes identifying when they occur, which pertains to offline analysis. For example, analyzing hospital admissions due to respiratory issues during a pandemic could involve detecting when new COVID-19 waves affect each hospital. If the number of change points is unknown, the problem involves determining the correct number of segments in the time series. Returning to the previous example, it involves discovering if and when new waves of COVID-19 appear. Over-detection may lead to false positives, while

E. Ogasawara et al., *Event Detection in Time Series*, Synthesis Lectures on Data Management, https://doi.org/10.1007/978-3-031-75941-3_4

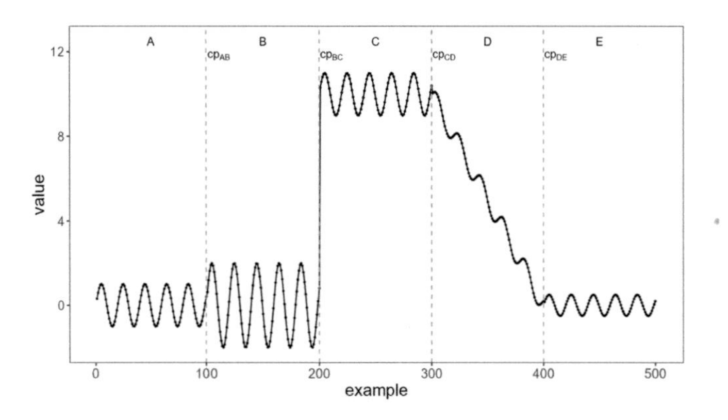

Fig. 4.1 A synthetic example of a time series with four change points

under-detection can result in false negatives. Addressing this dichotomy is an optimization problem aimed at distinguishing strong change points from weaker or spurious ones.

We have built a taxonomy for major change point detection methods, organized based on how change points are searched and representative detectors. It was adapted from Truong et al. [281]. In the taxonomy, the different types of search include segmentation, regression, and sliding windows (Fig. 4.2).

The search based on segmentation aims to minimize the distances among observations within each segment or maximize the distances between segments. When the number of change points (m) is known, a naive solution is to explore all possible segment partitions by brute force. If a segment is representative, the distances among its observations should be low, similar to clustering analysis. Different m values need to be explored when the number of segments is unknown. Increasing m continuously might reduce the distance of observations, potentially leading to each observation forming an individual segment. A penalty for each

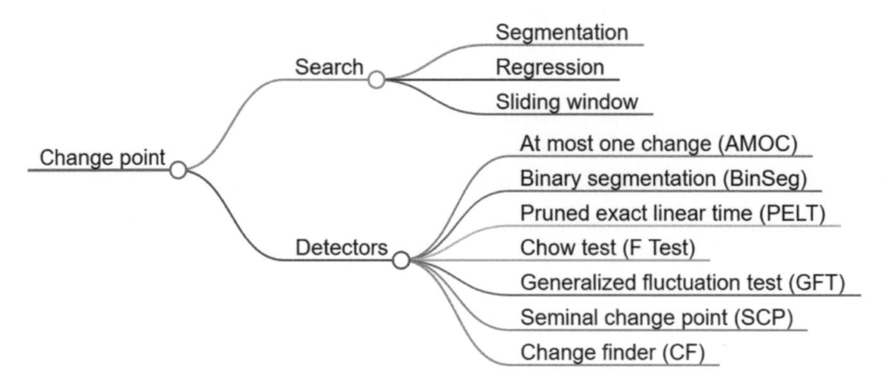

Fig. 4.2 Change point taxonomy (adapted from Truong et al. [281])

increase in the number of segments is used. Methods based on segmentation include At Most One Change (AMOC) [159] and Pruned Exact Linear Time (PELT) [160], detailed in Sect. 4.2.

The search for change points using regression involves exploring breakpoints in the time series by building a regression model for each candidate segment and measuring the quality of each segment using metrics like RSS. The lower the RSS, the better the established breakpoint. This leads to an optimization problem of finding the right breakpoints that minimize the RSS, usually supported by dynamic programming. Methods in this category include extensions of Chow Test (F Test) [25] and Generalized Fluctuation Test (GFT) [88], detailed in Sect. 4.2.

Finally, the search based on sliding windows involves computing all possible subsequences of a specified size p within the time series to detect change points. Inside each sequence, the goal is to check if a single change point exists. Many methods reside in this category, including some based on segmentation and regression [94]. Most sliding window methods search for a breakpoint and test whether the two segments within the window are significantly different. Methods in this category include Seminal Change Point (SCP) [123] and Change Finder (CF), detailed in Sect. 4.2. We have made representative change point detection methods available through our publicly Harbinger R package (see Appendix A).

4.2 Change Point Methods

This section explores the main methods, including AMOC [133], PELT [160], Chow Test [150], GFT [88], SCP [123], and CF [272]. At the end of this section, these detectors are compared for better comprehension using the synthetic time series presented in Fig. 4.1.

4.2.1 AMOC, BinSeg, and PELT

AMOC

AMOC focuses on finding a single change point in a time series X. Consider a time series X of size n, segmented into two parts at time k ($X_{1:k}$, $X_{k+1:n}$), where $1 < k < n - 1$. If there is a change at time k, the distributions of both segments differ from the entire time series X [159].

Each candidate segment is modeled using Maximum Likelihood Estimation with a specific PDF, such as Gaussian. The evaluation is a hypothesis test: the null hypothesis (H_0) indicates no change point, and the alternative hypothesis (H_1) suggests a change point at time k. A statistical likelihood test (assuming a normal distribution) evaluates a change in the mean [133]. The likelihood ratio method computes the maximum log-likelihood (`ML`) under the null (Eq. 4.1) and alternative (Eq. 4.2) hypotheses. In both equations, the $p(\cdot)$ function represents the probability density function of the sequences. For Eq. 4.1, θ_0 is the maximum likelihood of the parameters. Equation 4.2 provides the maximum likelihood from both sequences separated at time k, with estimated parameters (θ_1 and θ_2, respectively) [257].

$$\mathtt{ML}(X) = log\left(p(X_{1:n}|\hat{\theta_0})\right) \quad (4.1)$$

$$\mathtt{ML}(X, k) = log\left(p(X_{1:k}|\hat{\theta_1})\right) + log\left(p(X_{k+1:n}|\hat{\theta_2})\right) \quad (4.2)$$

From both Eqs. 4.1 and 4.2, it is possible to compute λ as described in Eq. 4.3, which gives the optimal k from all possible change point locations. Assuming a threshold cp, the null hypothesis is rejected when $\lambda > cp$. In this case, k is a change point for X. The challenge is to establish the value for the cp threshold.

$$\lambda = 2\left(\left(\arg\max_{k\in\{2,\ldots,n-1\}}\mathtt{ML}(X, k)\right) - \mathtt{ML}(X)\right) \quad (4.3)$$

The test proposed by Hinkley [133] is later expanded to support changes in variance, with a more general test combining mean and variance tests [159]. Besides, there are also alternatives to model the PDF that use different assumptions and parameters [317].

BinSeg

Binary Segmentation (BinSeg) is a method that uses a recursive and greedy algorithm on top of AMOC. The algorithm starts with an initial AMOC application and then runs AMOC on each subsequence produced from previous change point detections. This process is repeated until a stop criterion is met [281]. This greedy algorithm often provides a good approximation with a $O(nlogn)$ complexity.

PELT

The PELT method is an exact segmentation search using dynamic programming and some assumptions for efficiency [160]. The first assumption is that the number of change points increases linearly as the time series grows. The second assumption is that change points are spread throughout the time series rather than appearing close together. These assumptions result in a complexity of $O(n^2)$ [257].

4.2.2 Chow Test and GFD

Chow Test

The Chow Test test is driven by detecting structure change. It considers a time series X that can be modeled using linear regression, as described in Eq. 4.4, where y_t is the observation of the dependent variable, x_t represents the independent variables, and ω_t denotes white noise. The Chow Test test focuses on testing the null hypothesis of no structural change, i.e., $H0 : \beta_t = \beta_0$, against the alternative hypothesis that β_t varies over time [318, 319]. Several tests have been developed to evaluate this hypothesis, with F statistics-based tests particularly prominent. These tests, such as the $supF$, primarily target the alternative hypothesis of a single unknown breakpoint due to their ease of interpretation.

$$y_t = \beta_t x_t + \omega_t \tag{4.4}$$

The original idea for a structural change test was introduced by Chow [72], assuming that there is a change point. The method involves fitting separate regressions for two subsamples defined by the change point t_k and rejecting the null hypothesis when the F statistic exceeds a certain threshold [319]. Assume n is the number of time series observations, and r is the number of regressors in the linear model. An Ordinary Least Squares model is fitted for the observations before and after the potential change point t_k, estimating $2r$ parameters and computing the Error Sum Squared (ESS). Another Ordinary Least Squares model for all observations, with a RSS, is computed, estimating r parameters. The F statistic is given by Eq. 4.5 [319].

$$F = \frac{(RSS - ESS)}{ESS/(n - 2r)} \tag{4.5}$$

Methods based on the $supF$ test compute F statistics for all potential change points within an interval, defined by $t_a \leq t \leq t_b$, where t_i and t_f represent the first and last available time points, respectively. The null hypothesis of not having a change point is rejected when the F statistic becomes significantly high.

GFT

GFT is an alternative method for detecting change points based on regression, as defined by Eq. 4.4. It can identify multiple change points simultaneously, as it does not assume a specific deviation pattern from the null hypothesis. The method involves adjusting a model to the data and extracting an empirical process that captures the variations in residuals or parameter estimates. When no change point is detected (i.e., the null hypothesis holds), these variables adhere to central limit theorems, defining their boundaries with a fixed probability denoted as α. When a change point is detected (alternative hypothesis), the process's fluctuations generally increase [318].

As an example, consider an Ordinary Least Squares-based Cumulative Sum Control Chart (CUSUM) test, as introduced by Ploberger and Krämer [229]. This test relies on cumulative sums of standard Ordinary Least Squares residuals (ω_i), defined by Eq. 4.6. In this context, $W(\cdot)$ represents standard Brownian motion. The limiting process of $W_n^0(u)$ resembles the standard Brownian bridge $W^0(u) = W(u) - uW(1)$. Under a single-shift alternative, this process is expected to peak around the change point [318, 319].

$$W_n^0(u) = \frac{1}{\hat{\sigma}\sqrt{n}} \sum_{i=1}^{\lfloor nu \rfloor} \hat{\omega}_i, \; (0 \leq u \leq 1) \tag{4.6}$$

4.2.3 Seminal Change Point and Change Finder

SCP

The SCP for detecting change points has become a reference in the literature [123] for sliding window-based methods. For a time point in a sliding window, models are adjusted to segments before and after that point. A change point is determined if the total fitting errors are significantly reduced compared to when no change point exists. In its simplest version, the method uses linear regression to adjust the data, with the adjustment error measured in terms of square errors [272].

Figure 4.3 illustrates how the method works. The blue line represents the regression for the entire window, with the candidate change point observation marked in gray. The regressions for observations before and after the candidate observation are marked in green and red, respectively. If the sum of residuals for all observations concerning the blue regression is significantly higher than that of residuals from before and after regressions, then the candidate observation is, in fact, a change point.

Change Finder

The creation of SCP has enabled the development of many other sliding window-based methods, such as CF [272]. CF consists of two phases. In the first phase, given a time series X, a model α is adjusted using an autoregressive model, such as AR. Let $\omega_i = (x_i - \hat{x}_i)^2$, which represents the squared residual from the model. Outliers in this residual series W are directly marked as anomalies.

The insight of CF is to compute a moving average for the residual series W, represented as the Z time series. This procedure provides inertia for the residuals. Z is modeled again using an autoregressive model, such as AR, resulting in the $\hat{Z}$ series. This produces a residual series R, where $r_i = z_i - \hat{z}_i$. Outliers in R are considered change points.

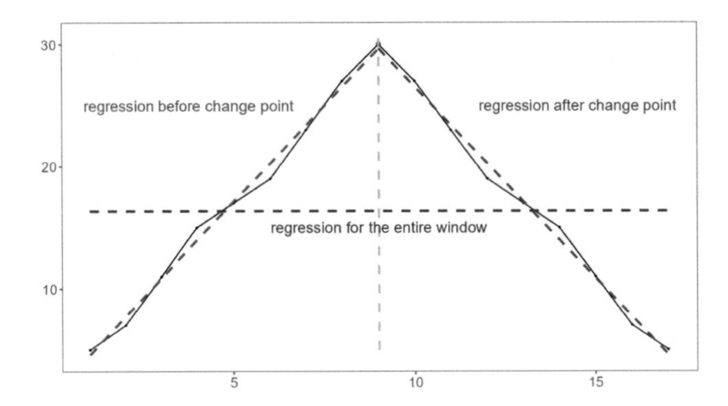

Fig. 4.3 Detection strategy of SCP

4.2.4 Comparison of Change Point Detectors

Figure 4.4 shows the synthetic time series from Fig. 4.1 with detections made by AMOC, BinSeg, PELT, Chow Test, GFT, SCP, and CF using ARIMA. Green marks indicate correct detections (true positives), blue marks indicate undetected events (false negatives), and red marks indicate incorrect detections (false positives).

AMOC detects changes in mean and variance, making a single detection for the time series, which occurs close to time 400. This makes sense as the mean of the prior segment is higher than the mean after the detection, which is close to zero. BinSeg makes a recursive application of AMOC for each segment (before and after detection) when it makes sense. The first change point detection is close to the actual change point, and the second is correctly detected, but it detects many points in a downtrend between 300 and 400. PELT, similar to BinSeg, detects 11 change points, with many detections between 300 and 400 due to the constant change in the mean. Chow Test detects a single change point, driven by structural

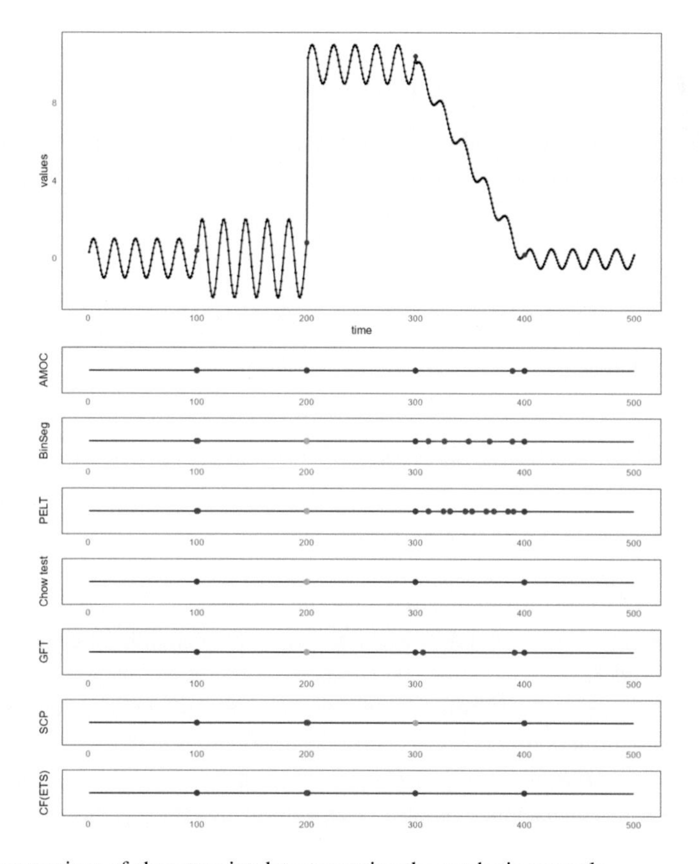

Fig. 4.4 A comparison of change point detectors using the synthetic example

changes in the time series, identifying the change close to 200 due to the significant trend change. GFT, targeting trend changes, makes the correct detection at time 200 and two detections close to times 300 and 400, though it misses the first variance change. SCP using a sliding window of 60 makes two detections: a correct one at time 300 and a close one at time 200. The method is sensitive to the sliding window size. Using the ETS model, the CF detects a single change point close to time 200, being sensitive to the adopted method and sliding window size.

4.3 Concept Drift

Concept drift is the online monitoring of changes that occur in a streaming time series. Consider a multivariate time series D containing independent variables (X) and a target variable (Y), where D is organized into batches with a varying distribution χ, relating $X \rightarrow Y$. A concept for D at a batch i (SD_i) is defined as the probability of the distribution χ for D, described in Eq. 4.7.

$$concept(SD_i) = p(\chi_i) \tag{4.7}$$

A concept drift between batches i and $i+1$ is defined as a statistically significant difference between the probabilities $p(\chi_i)$ and $p(\chi_{i+1})$, described in Eq. 4.8 [294].

$$p(\chi_i) \neq p(\chi_{i+1}) \tag{4.8}$$

We have built a taxonomy for concept drift, organized by category and type of drift. It is presented in Fig. 4.5. Detectors are organized by type and method, and drift handling approaches are included.

4.3.1 Category and Types of Drifts

Drifts are categorized as virtual or real. Virtual drifts involve changes in the distribution of X or Y without changing the concept χ. Real drifts involve changes in the concept χ, with or without changes in the distribution of X or Y. For example, if a model M is built from X to predict Y $(M(X) \rightarrow \hat{Y})$, a real drift is characterized by a significant change in the error rate between Y and $\hat{Y}$.

Concept drifts are also classified as abrupt, incremental, gradual, and reoccurring. Figure 4.6 illustrates these four types of concept drift. The time series is organized into four batches $(SD_1, \ldots, SD_4)$ of size eight. Observations marked in red correspond to change points, while dashed black lines indicate a change in the concept between consecutive batches.

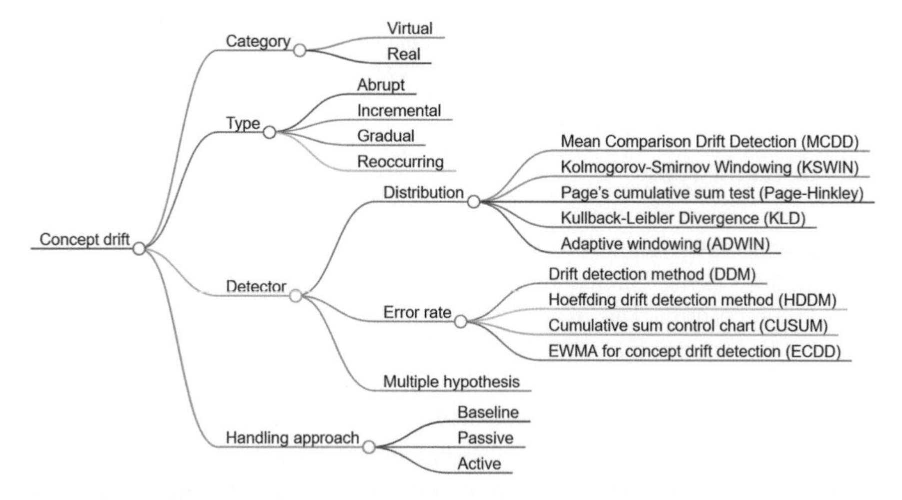

Fig. 4.5 Taxonomy of concept drift, including the type of drift, drift detector type, detector method, and drift handling approach

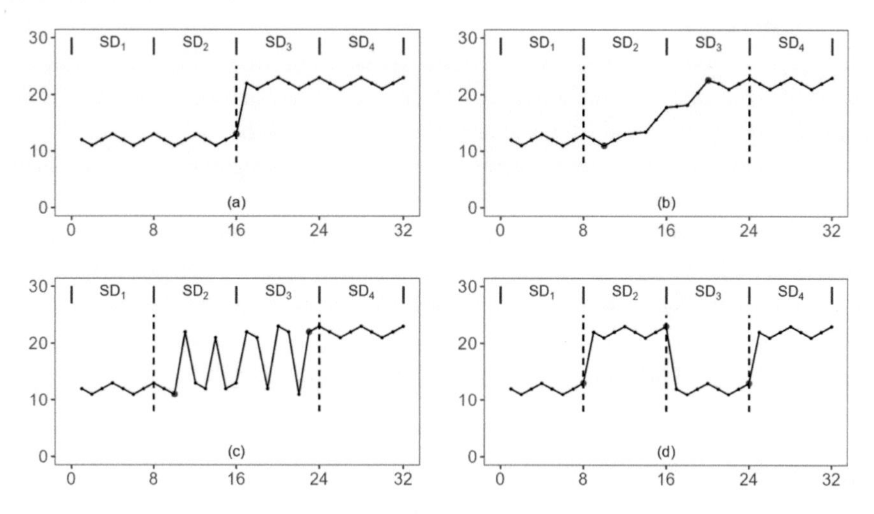

Fig. 4.6 Different types of concept drift: abrupt (**a**), incremental (**b**), gradual (**c**), and reoccurring (**d**) (adapted from Bayram et al. [35])

In an abrupt drift, the concept changes significantly in the following batches, as shown in Fig. 4.6a from batch SD_2 to SD_3. Incremental drift occurs smoothly, with observations transitioning from one concept to another, as depicted in Fig. 4.6b from batch SD_2 to SD_4.

Gradual drift, shown in Fig. 4.6c, involves the concept intermittently changing from one concept to another until it stabilizes at the new concept. The concept starts changing at batch SD_2 but only stabilizes at the new concept at batch SD_4. Reoccurring drift, illustrated in

Fig. 4.6d, depicts periodic behavior where the concept switches back to the previous one, occurring between batches SD_1 to SD_2, SD_2 to SD_3, and later from SD_3 to SD_4.

4.3.2 Detector Type

Changes in time series distribution may result in having more events in the time series when the concept drifts occur. Furthermore, event detection models might be vulnerable to the presence of concept drifts. Thus, detecting and handling drifts is a relevant subject associated with event detection, enabling a specific action to avoid increasing errors in online learning systems after a drift is observed. There are three main categories of drift detection [35]: data distribution, error rate, and multiple hypothesis tests.

Data distribution-based methods use statistical inference and analysis of feature distribution to detect significant output class proportion changes concerning input variables. Once a new batch arrives, the distribution of the batch is compared with the distribution of the data used to train the models [142]. They are connected to change point detection methods based on distribution.

Error rate methods use a learning algorithm (statistical or ML) [145] and indicate a drift based on the error rate of prediction results. Once a new batch arrives, an inference is made. The predictions are used and later evaluated to determine whether they are correct. A drift is assumed to exist once the error obtained from this batch is significantly higher than the previously observed errors [106].

Finally, the multiple hypothesis tests are based on data distribution and error rate methods. Thus, they are built as an ensemble of methods, i.e., they can be parallel-checked, and the most voted result indicates the presence of a drift. Alternatively, the test can be structured as a stack. A fast detector with more chance of having false positives triggers another detector, which is more time-consuming but potentially more accurate [183].

4.3.3 Handling Approach

The handling approach establishes how models are adjusted as time evolves as a way to address concept drift. Consider a multivariate time series (D) partitioned into batches of size b. SD_1 and SD_n correspond to the first and last batches of D, respectively. Yet, a batch of size b at time i is formally defined as $SD_i = < D_{(i-1) \cdot b+1}, \ldots, D_{i \cdot b} >$, such that $|SD_i| = b$. It can target the detection and handling of concept drift.

From these concepts, it is possible to define three approaches to handle concept drift: (i) *baseline*; (ii) *passive*; (iii) *active*. In the *baseline* approach, a model is built using the first batch. The trained model is continuously used. No action is taken when drift occurs, and the trained model might increase its error when inferencing newer batches. It corresponds to Fig. 4.7a, where the first batch (SD_1) is used for training a model (indicated as a red square) for inferencing all other batches (SD_2 to SD_n).

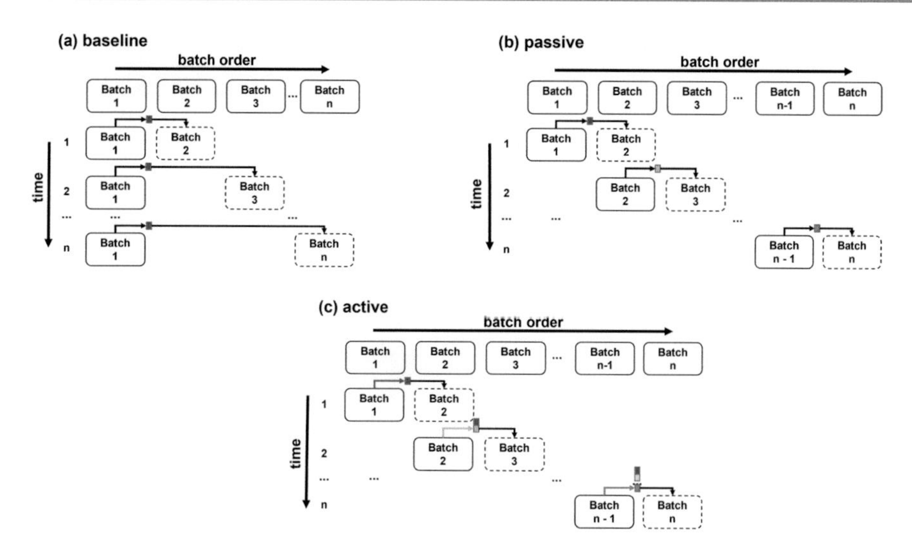

Fig. 4.7 Drift handling approaches. **a**: *baseline*; **b**: *passive*; **c**: *active*

In a *passive* approach, it is assumed that drift always occurs. Thus, batch SD_i is used for training the model to inference batch SD_{i+1}. This scenario is shown in Fig. 4.7b, where models are constantly updated (they are presented in different colors). The drawback of this approach is that it might retrain models, even if no drift occurred in the time series [107].

Finally, the *active* approach applies drift detection whenever a new batch is introduced. If no drift is detected, the previously trained model is still used. However, when drift occurs, a new model is built using the previous batches. Figure 4.7c depicts this scenario. A new model (in orange) is used if drift occurs between batches two and three. Otherwise, the previous model (presented in red) is preserved. In this approach, two extreme scenarios may occur. The first is that the same model can be used from the first batch to the last one, resembling the baseline approach. The difference is that a model is only kept if no drift has been detected. The second is continuously retraining in all batches, resembling the *passive* approach. Again, such a decision is based on whether drift is observed whenever a new batch is introduced [142]. Moreover, some methods may change the size of the batches using soft computing (fuzzy systems) to identify drifts and adapt accordingly [184].

4.4 Concept Drift Methods

This section covers the main methods within the broad range of concept drift detectors [126, 183]. Figure 4.8 shows the overall framework for drift detection, which encompasses four steps: data retrieval, modeling, dissimilarity measure, and hypothesis test. The data retrieval step establishes how data is collected from streaming and passed to drift detection. The data

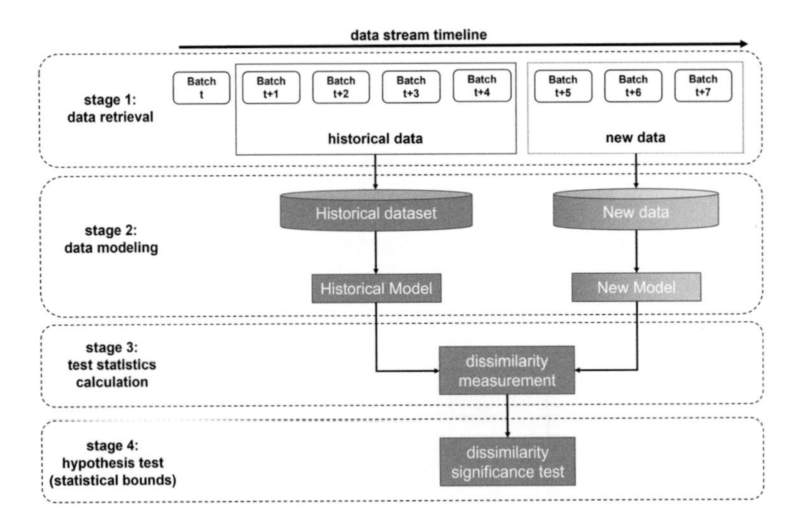

Fig. 4.8 General framework for drift detection (adapted from [183])

modeling targets creating a model to support drift detection. It might demand extracting features and conducting transformations that might increase the performance or accuracy of drift detectors. The dissimilarity measures aim to measure the difference between previous and new batches. It quantifies the severity of the drift to support the hypothesis test. The last step evaluates the statistical significance of the change observed in the dissimilarity step to characterize a drift. We have made representative concept drift detection methods available through our publicly Heimdall R package,[1] a framework for concept drift detection.

The methods are organized into two groups: error rate and distribution. The former includes Drift Detection Method (DDM), Hoeffding Drift Detection Method (HDDM), CUSUM, and EWMA For Concept Drift Detection (ECDD). They are driven to detect real drifts, as the relationship between predictors and predictand changes. The latter includes Mean Comparison Drift Detection (MCDD), Kullback–Leibler Divergence (KLD), Kolmogorov–Smirnov Windowing (KSWIN), Page's Cumulative Sum Test (Page-Hinkley), and Adaptive Windowing (ADWIN). This group analyzes changes in the distribution of predictors. To this extent, they are, by definition, targeting virtual drifts.

4.4.1 DDM and HDDM

DDM

The DDM uses a binomial distribution to model the number of classification errors in a batch b_t. DDM calculates, for each instance i in the data stream, the probability of misclassification

[1] https://cran.r-project.org/web/packages/heimdall/index.html.

p_i and its standard deviation σ_i. If the distribution of samples is stationary, the model maintains its performance [111]. Suppose the error rate of the learning algorithm increases significantly. In that case, it suggests changes in the class distribution, indicating that the current model is inconsistent with the current data, thus providing a signal of concept drift.

While monitoring a time series, DDM calculates the values of p_i for each instance. When $p_i + \sigma_i$ reaches its minimum value, it stores $p_{\min}$ and $\sigma_{\min}$. Then, DDM checks two conditions to detect whether the system is in the alert or concept drift level: the alert level is triggered when $p_i + \sigma_i \geq p_{\min} + 2\sigma_{\min}$. Beyond this level, it stores instances anticipating a possible concept drift. The concept drift level is triggered when $p_i + \sigma_i \geq p_{\min} + 3\sigma_{\min}$. At this level, DDM resets the variables ($p_{\min}$ and $\sigma_{\min}$) and the induced model. Subsequently, a new model is built using the instances stored since the alert level has been triggered.

DDM performs well in detecting gradual and abrupt changes but struggles to detect drifts in incremental scenarios. In this case, storing samples can be accumulated over time before the concept drift level is triggered [111].

HDDM

The HDDM [103], like DDM, Fast Hoeffding Drift Detection Method (FHDDM) [224], and McDiarmid Drift Detection Method [225], belongs to a group of methods that include drift detection methods based on prediction error analysis. The HDDM method monitors an estimated performance based on classifier errors using Hoeffding's inequality. It adapts step four of the general framework for drift detection (depicted in Fig. 4.8) to introduce Hoeffding's inequality.

Consider a batch b of observations x_i, such that $x_i \in [l_i, h_i]$, with $i \in \{1, \ldots, |b|\}$. Consider also $\overline{X}$ as the computed mean of observations. For any ϵ threshold ($\epsilon > 0$), Eq. 4.9 indicates a boundary for the probability of the computed mean being different from the expected mean ($E[\overline{X}]$, computed from previous batches) by a ϵ threshold.

$$Pr\{\overline{X} - E[\overline{X}] \geq \epsilon\} \leq e^{-\frac{2\epsilon^2}{\sum_{i=1}^{n}(h_i - l_i)^2}} \tag{4.9}$$

Equation 4.10 enables estimating the error ϵ for a significance level of δ. HDDM establishes bounds to check the distance or inequality between the error distributions during predictions. In other words, the difference between the number of false positives or false negatives at batch i is compared to those in a measurement taken at $i + 1$ to indicate concept drift.

$$\epsilon_\delta = \sqrt{\frac{1}{2n} ln\left(\frac{1}{\delta}\right)}. \tag{4.10}$$

4.4.2 CUSUM

CUSUM is a sequential analysis method [107], designed to detect changes in distribution using a criterion for deciding when to take corrective action. As the name suggests, the CUSUM involves the calculation of a cumulative sum (hence the term sequential) time series Z. Considering the general framework, the hypothesis test is verified when Z significantly deviates from zero.

The time series Z uses an additional quality time series (ω). An example of ω is the residual, i.e., the difference between the model and time series. In this case, observations from a time series z_t are assigned from previous values z_{t-1}, ω_t, and a magnitude parameter δ. It is described in Eq. 4.11. When the value of z_t exceeds a certain predefined threshold, it is identified as a change point, characterizing a concept drift.

$$z_t = \begin{cases} 0, & t = 1 \\ \max\left(0, z_{t-1} + \omega_t - \delta\right), & t > 1 \end{cases} \tag{4.11}$$

For problems of binary classification (having or not an event), δ can be set to zero in Eq. 4.11 and ω_t should be 1, in case of an error, and -1, otherwise.

4.4.3 ECDD and MCDD

ECDD

ECDD applies an inertia-based drift detection. It uses the Exponentially Weighted Moving Average (EWMA) as an inertial follower adopted for monitoring time series. The exponential property progressively gives less weight to older data points as they move back in time. A parameter λ determines the rate at which older data enters the statistic. EWMA is represented in Eq. 4.12 as z_t. From the properties of EWMA, the mean and standard deviation for the inertial time series Z at a time t are characterized in Eq. 4.13. The μ_{z_t} corresponds to the time series mean (μ_{X_t}) at time t [242]. The EWMA is a preprocessing step for the general framework depicted in Fig. 4.8.

$$z_t = \begin{cases} \mu_t, & t = 1 \\ \lambda \cdot \mu_t + (1 - \lambda) \cdot \mu_{t-1}, & t > 1 \end{cases} \tag{4.12}$$

$$\mu_{z_t} = \mu_{X_t}, \sigma_{z_t} = \sqrt{\frac{\lambda}{2 - \lambda}(1 - (1 - \lambda)^{2t})\sigma_{X_t}} \tag{4.13}$$

In ECDD, before a change point $\mu_t = \mu_a$, the time series varies around the follower z_t. When a change occurs, the value of μ_t changes to μ_b and z_t gradually moves toward μ_b. In this case, it is possible to establish a control limit L to determine how much z_t must diverge from μ_a before characterizing a concept drift. Such limit is established by Eq. 4.14.

$$z_t > \mu_a + L \cdot \sigma_{z_t} \tag{4.14}$$

MCDD

The MCDD method compares the mean and variance of a given predictor variable a in previous batches (B_{t-1}) and the current batch (b_t). It determines whether to use a parametric or nonparametric test based on the distribution of the samples. Considering a memory size m, $B_{t-1}(a) = \{b_{t-m+1}(a), \ldots, b_{t-1}(a)\}$ is the set of previous batches for the variable a. Considering also that the current batch for the variable a is $b_t(a)$, if the expected value for $b_t(a)$ is significantly different than the expected set of batches $B_{t-1}(a)$ (i.e., $E(b_t) \neq E(B_{t-1})$), a drift is detected. Otherwise, no drift is observed [110].

4.4.4 Page-Hinkley and KSWIN

Page-Hinkley

The Page-Hinkley test is a variation of CUSUM [107]. It considers a cumulative time series Z expressed as z_t, accumulating the difference between the values x_t and the mean of X up to time t ($\overline{x}_t$). Equation 4.15 characterizes the Page-Hinkley test (PH), where $\overline{x}_t = \frac{\sum_{i=1}^{t} x_i}{t}$ and δ correspond to the magnitude of the allowable changes.

$$z_t = \sum_{i=1}^{t} (x_i - \bar{x}_i - \delta) \tag{4.15}$$

This difference signals a drift when it exceeds the threshold δ, which depends on the acceptable false alarm rate. Increasing δ results in fewer false alarms (false positives) but may lead to missed (false negatives) or delayed change detection.

KSWIN

Considering two consecutive batches (b_{t-1}, b_t), the KSWIN test is a nonparametric statistical test that compares the distributions of two data samples to determine if they differ significantly. The KS test statistic is the maximum absolute difference between the Cumulative Distribution Function (CDF)s of the two samples.

The statistical test adopts the null hypothesis (H_0) that the two batches come from the same distribution. The alternative hypothesis (H_1) establishes that the two batches come from different distributions. Suppose the maximum absolute difference between the two empirical CDFs exceeds the critical value at a significance level p-value (often set at 0.05). The null hypothesis is refuted, indicating a drift in the data distribution [236].

4.4.5 KLD

The KLD is a drift detection method that compares the probability distributions of two samples: older and most recent. Given two probability distributions P (older) and Q (most recent) of a discrete random variable X that presented in a batch with b buckets, $\{x_1, x_2, \ldots, x_b\}$, where b is the number of buckets, the KLD is defined as Eq. 4.16, where P and Q are the probability distributions of the two samples.

$$D_{KL}(P(X)||Q(X)) = \sum_{i=1}^{B} P(x_i) \cdot ln\left(\frac{P(x_i)}{Q(x_i)}\right) \tag{4.16}$$

The KLD can be understood as a measure of the expected excess surprise using the actual distribution versus the expected distribution as a divergence. To apply this method to detect drifts, the algorithm keeps two batches (present and past) and calculates the KLD each time new data is included, and the state is updated. The method indicates that drift occurs whenever the KLD is above a specific threshold (such as 0.9).

4.4.6 ADWIN

ADWIN [38] uses variable-sized sliding windows, recomputed online based on the observed rate of change in the data within these windows. The window (W) dynamically expands when there is no clear change in the context and contracts when a change is detected. If a change is identified, it eliminates an outdated portion of the window.

ADWIN relies on the use of Hoeffding's bound [37] to detect drift, comparing the average difference between two sub-windows ($W0$ for older instances and $W1$ for recent instances) within W. If the average difference is statistically significant, then ADWIN removes all instances from $W0$ that are considered to represent the old concept and only retains $W1$ for the next test.

4.4.7 Comparison of Drift Detectors

To better understand the methods, let us compare them using the example of Fig. 4.1. The time series plots correspond to the predictor variable X. Let us assume a response variable Y, such that $y_i = true$ whenever x_i is greater than four and $y_i = false$ otherwise. Let us also introduce a naive prediction model $M(X)$ that predicts all values as negative $\hat{y}_i = true, \forall i$. This model is right when time series values are below four and incorrect otherwise. In this case, drift detection is evaluated whenever an observation is processed.

Figure 4.9 shows the detections made by the main methods previously presented, which are organized in two scenarios. The methods based on error rate evaluate the model errors (DDM, HDDM, CUSUM, ECDD) as each observation arrives. The methods based on distri-

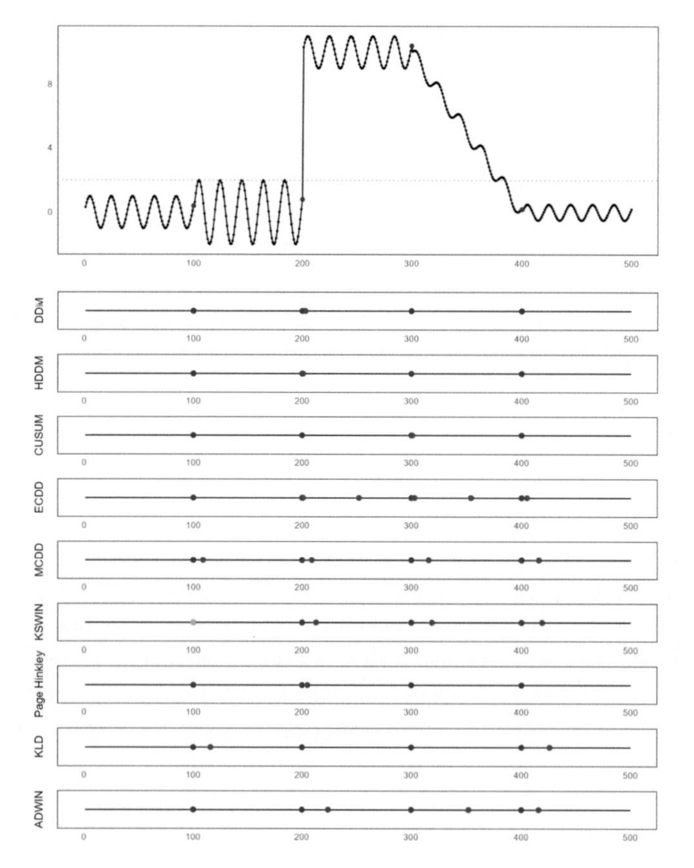

Fig. 4.9 Concept drift detection using DDM

bution evaluation (MCDD, KLD, KSWIN, Page-Hinkley, and ADWIN) analyze X to check for changes in distribution during batch processing.

Regarding the error rate methods, the DDM detects drift close to time 200 when the prediction model significantly increases the error. No drift is observed when the error decreases at time 400. The HDDM, similarly to DDM, detects drift at time 200 when the prediction model increases the error significantly. However, it is also detected between 300 and 400 since the error decreased in time 400. The CUSUM makes a detection close to time 300 when the error rate is close to maximum. Finally, the ECDD detects change points regularly when the model error increases (at a time close to 200) until it recovers a low error (at a time close to 400).

Regarding the distribution-based methods, the MCDD detects drifts relatively close to the labeled concept drift in the time series. The KSWIN behaves very similar to MCDD. The detection of KSWIN is usually slightly more distant than MCDD. Nevertheless, the KSWIN detects the first drift at the right moment (time 100). As expected, the Page-Hinkley can only detect abrupt changes in mean. The drift is discovered close to time 200. Using KLD, it is possible to observe that the first drift is detected in the presence of volatility change. The second one is close to the intervals where the distribution gets stabilized. Finally, the ADWIN cannot detect the volatility change at time 100 but reacts to other drifts with a certain delay.

Benchmarking drift detector methods has received attention recently, with works exploring various methods (real and virtual) and types of drifts (gradual, abrupt, and incremental) [244]. Some general observations can be drawn using a broad comparison, but the choice of the best algorithm depends on the specific requirements and the type of drift to be detected. Furthermore, the general behavior might not reproduce directly in specific domains, as it can be observed in flight [110], IoT [1], and security [116].

4.5 Advanced Topics

4.5.1 Multivariate and Spatial–Temporal Time Series

The number of dimensions can increase the difficulty of detecting change points. Some recent works focus on detecting change points in multivariate time series. For instance, Prabuchandran et al. [232] model compositional time series with a parametric Dirichlet distribution, which allows identifying multiple change points by iteratively detecting single change points within a sliding window and using a permutation test to assess their significance. Similarly, Hlávka et al. [134] use empirical characteristic functions to detect changes in pairs of multivariate time series. Their method projects vector observations into a one-dimensional space and uses distance-based criteria and resampling procedures to detect changes in online and offline scenarios.

Like multivariate time series, spatial–temporal time series require sophisticated methods for detecting drifts and change points. Several studies have focused on addressing these challenges. Amador Coelho et al. [22] use quadtrees to monitor the spatial distribution of data in the feature space. A separate quadtree represents each class in a binary classification problem, and misclassified data triggers an inspection of the corresponding quadtree to detect concept drift. Similarly, Liu et al. [178] focus on detecting regional density changes in data streams, introducing a local drift degree (LDD) metric to measure the likelihood of regional drifts. They use a regional drift adaptation algorithm (LDD-DSDA) to merge existing training data with recently buffered data. Their method synchronizes density discrepancies based on LDD measurements with non-drifted data, reducing the risk of overestimating drift regions. Chen et al. [68] propose a change point detection method that combines multinomial logit models with multidimensional latent Dirichlet allocation (LDA) to decompose attributes into a mixture of activities. A Dirichlet multinomial regression (DMR) accounts for the changing regularity of activity prevalence. Their method is evaluated using bike-sharing data in New York City, capturing both spatiotemporal activities (departure days, times, and destinations) of commuting behavior and significant changes over time.

Some works are targeted to specific domains such as sports and industry. Kim et al. [162] introduce SoccerCPD, a spatial–temporal change point detection method specifically designed for analyzing soccer matches. This method differentiates between tactical formation changes and role changes in teams, distinguishing them from temporary adjustments made during the game. Florez et al. [102] introduce CatSight, a method for detecting change points in spatial–temporal multivariate time series tailored for industrial monitoring. CatSight uses the Common Spatial Pattern method to project multivariate temporal data into a new space that maximizes the variance difference between classes before and after drift. This transformation enables the extraction of the most discriminative features, which are then used by conventional ML classifiers to detect changes in the data stream.

4.5.2 Spectral-Based Methods

In recent years, spectral-based methods have gained attention for concept drift and change point detection, including PCA and autoencoders. They add value to address the high dimensions of multivariate time series. PCA-based methods provide dimensionality reduction to support concept drift detection [1]. Autoencoders have been recently explored to support concept drift detection. Some works are based on reconstruction error analysis, including [169], where the Mann–Whitney U Test compares reconstruction losses between a reference window and recent data. Upon detecting a drift, the autoencoder is retrained with new data to maintain performance in nonstationary environments. Other methods are based on the latent space of autoencoders [171]. While it detects anomalies based on reconstruction errors, concept drift is detected based on changes in the latent space using statistical tests.

Recent works also utilize autoencoders for change point detection in time series. De Ryck et al. [84] detect change points by learning time-invariant features that remain stable within segments without change points. Their method preprocesses time series with discrete Fourier transforms to capture time-domain and frequency-domain information and trains autoencoders to distinguish between time-invariant and instantaneous features. Change points are identified using a dissimilarity measure based on the Euclidean distance between time-invariant features of consecutive windows. Gupta et al. [122] develop a three-phase architecture for real-time change point detection: Deep Adaptive Input Normalization (DAIN) to normalize the data adaptively, Recursive Singular Spectrum Analysis (SSA) to decompose and smooth the data, and an autoencoder to detect change points by analyzing reconstruction errors. The errors are compared against a dynamically adjusted threshold to identify significant mean, variance, frequency, and autocorrelation changes. Both methods ensure accurate and flexible detection of change points in nonstationary time series.

4.6 Further Reading

Several surveys and books on change points are available. Aminikhanghahi and Cook [23] present a survey for change point detection methods and Truong et al. [281] review methods for multivariate time series change points. Additionally, Chen and Gupta [66] present parametric methods for detecting change points. Sayed-Mouchaweh et al. [251] comprehensively review recent developments and methodologies associated with building models in nonstationary scenarios.

Baburoglu et al. [33] provide a bibliometric review of concept drift from 1980 to the present, offering a comprehensive view of the subject. Hoens et al. [135] present approaches for handling concept drift in imbalanced classification problems. Gama et al. [107] and Khamassi et al. [157] review adaptive learning processes in data stream scenarios. Pratama et al. [233] present incremental learning methods using recurrent fuzzy neural networks. Bayram et al. [35] review model degradation under concept drift, and Han et al. [126] compare passive versus active approaches for tackling concept drift.

Despite the prevalence of deep learning in building prediction models, addressing concept drift poses challenges such as computational cost and the absence of new data to adapt deep learning models. Xiang et al. [302] summarize concept drift adaptation methods under the deep learning framework. Several studies focus on time series, including You et al. [314] and Herbert et al. [132], which discuss orchestrating training and updating recurrent models under concept drift scenarios. Yuan et al. [316] present concept drift adaptation for abrupt, gradual, and recurrent types of drift. Wang et al. [292] evaluated transfer learning to support deep learning adaptability to concept drift. Finally, we have proposed the DJEnsemble model selection approach for reacting to variations in the time series distribution. The approach adapts to time series concept drift by selecting an ensemble of pre-trained models whose training data distribution mostly resembles the input time series distribution [223].

4.7 Conclusion

This chapter presented the fundamentals of change points, characterized by changes in time series dynamics (distribution or autocorrelation). It also covered the difference between offline and online analysis, with offline focusing on identifying past changes and online aiming to detect changes as early as possible.

The chapter explored several representative change point detection methods, such as AMOC, BinSeg, PELT, Chow Test, GFT, SCP, and CF. Their usage in a synthetic example helps to illustrate the properties in detecting various changes, providing insights into their practical applications and limitations.

The chapter also covered concept drifts, which are changes in the relationship between independent and dependent variables over time. Drifts are classified as virtual or real and organized based on their occurrences, which are abrupt, incremental, gradual, and reoccurring. Drift detection methods are organized based on data distribution, error rate, or multiple hypothesis tests. The chapter also presented the main drift handling approaches, including baseline (no action), passive (continuous model updating), and active (model updating upon detecting drift). Main drift detection methods are presented, including DDM, HDDM, CUSUM, ECDD, MCDD, KSWIN, Page-Hinkley, and ADWIN.

Advanced topics ended this chapter by covering multivariate and spatial–temporal time series. As a trend, many novel methods, especially those based on spectral methods, are expected to support multivariate, spatial–temporal, trajectory, and streaming time series. Novel methods might also adopt a high level of time series preprocessing to support non-stationarity.

Motif Discovery

5

5.1 Motifs

Motif discovery identifies recurring sequences in a time series, where the repeated subsequence is initially unknown and discovered through scanning. Motif discovery has broad applications, including anomaly detection, where identifying recurring patterns highlights deviations from the typical observations are often signaled as anomalies [282].

In financial time series analysis, motif discovery informs trading strategies and risk management by uncovering patterns in vast time series, such as stock prices and trading volumes [285], whereas in bioinformatics, motif discovery is essential for identifying repeated sequences in genomic sequences [258]. Similarly, in environmental monitoring, detecting recurring patterns in sensor data related to temperature, humidity, or pollution levels helps understand environmental changes and address issues such as climate change [230].

In this context, we formalize and detail the general concepts related to motif discovery. Still, before formalizing it, we present the concept of sequence occurrences in a time series, from which the definition of a motif is derived. Let q be a sequence of size p. Consider W as the set of sliding windows of size p obtained from a time series X. As described in Eq. 5.1, the set of occurrences of q in X is the set of windows w_i of W that are similar to q [252].

$$occurs(q, X) = \{w_i\} \mid \forall w_i \in W, w_i \approx q \quad (5.1)$$

The sequence q is a *motif* in X with support σ if and only if q occurs in X at least σ times, as described in Eq. 5.2. The length p of a motif q is also known as word size [44].

$$motif(q, X) \iff |occurs(q, X)| \geq \sigma \quad (5.2)$$

Some definitions of motifs require that their occurrences should not be trivial. The sequences in $occurs(q, X)$ should have minimal overlaps regarding the time they occur,

E. Ogasawara et al., *Event Detection in Time Series*, Synthesis Lectures on Data Management, https://doi.org/10.1007/978-3-031-75941-3_5

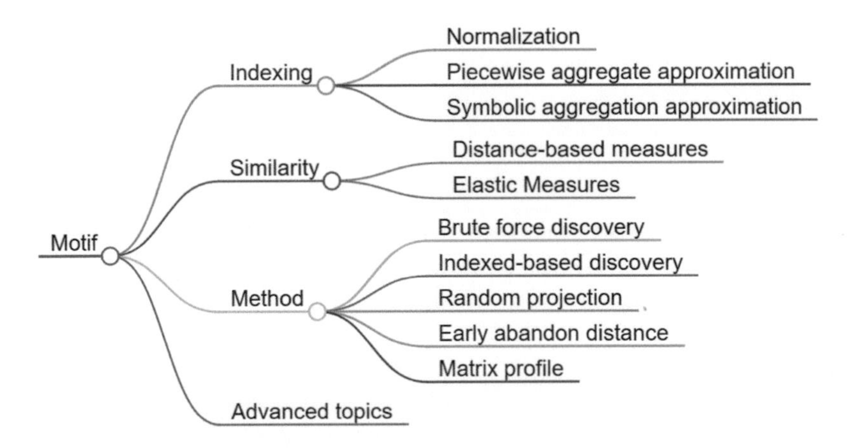

Fig. 5.1 Basic taxonomy for motif discovery

usually achieved using exclusion zones (between 50 and 80%) among them [199]. The most significant motif (1-motif) in X is the most occurred subsequence, disregarding trivial matches formed by overlapping subsequences. Consequently, the k-motif is the kth most occurred motif in the time series [176]. Another important aspect is the motif's significance in a time series. If a motif q occurs too frequently, it loses the expected property of being previously unknown. Initially, such subsequences are characterized as motifs but may be classified as false motifs if they have little relevance and offer minimal or no useful information [71].

We have organized the main concepts influencing motif discovery in a taxonomy (see Fig. 5.1) that includes time series indexing, similarity functions, and discovery methods, all described in the following sections.

5.2 Time Series Indexing

Many proposed methods for discovering motifs in time series are computationally intensive [221]. Therefore, various time series indexing techniques are used to improve the effectiveness of motif discovery methods in reducing computational resources. These techniques include data preprocessing, such as normalization and encoding, to enhance the performance and precision of the results [199].

Normalization

Certain properties like scale must be verified before similarity searches, making normalization an important step in data preprocessing [156]. A detailed description of various normalization techniques is presented in Chap. 2 and summarized as follows.

One of the most widely used normalization techniques is the Z-Score, which results in a time series with a mean of zero and a variance of one. Other normalization techniques include Min-Max, unit length, and mean normalization. The Min-Max scales data to a specified range (e.g., [0, 1]) [210]. Unit length normalization scales data points so that the entire time series has a length of one, while mean normalization adjusts data by subtracting the mean and dividing by the range [217].

Advanced normalization techniques, such as Adaptive Scaling and Adaptive Normalization, introduce resilience to nonstationarity. Adaptive Scaling computes a scaling factor between pairs of time series to improve their comparability, while Adaptive Normalization dynamically adjusts the scaling factor based on local characteristics of the time series, enhancing comparability between different time series [209]. Each method addresses specific distortions that may aid the motif discovery process [217].

Piecewise Aggregate Approximation

Piecewise Aggregate Approximation (PAA) is a temporal aggregation technique that reduces the number of observations in a time series. Given a time series X with n observations, the number of observations is reduced to n', resulting in X'. Each observation in X' is the mean of equally time observations concerning X, where X is partitioned into k segments, such that $k = \frac{n}{n'}$. Equation 5.3 characterizes this transformation [305]. The overall complexity of PAA is $O(n + k)$.

$$x'_j = \sum_{t=n'(j-1)+1}^{n' \cdot j} \frac{x_t}{n'} \tag{5.3}$$

Figure 5.2a shows a fraction of patient #102 heartbeat in an electrocardiogram[1] normalized with Z-Score. Figure 5.2b illustrates PAA applied to that patient's heartbeat using $n' = 20$.

Symbolic Aggregation Approximation

Time series observations are typically real values, and processing them directly in numerical representation may not be efficient for motif discovery [82]. Discretization methods, such as SAX, represent a range of values as symbols, enabling the technique to support exact match search to speed up motif discovery. SAX partitions the time series into ranges, each associated with a particular symbol [176]. The SAX alphabet size determines the number of partitions for the domain, and all values are replaced by their respective associated symbol. Given a time series X with n observations and an alphabet $(a_1, \ldots, a_m)$ of size m, the values of X are divided into m ranges (e.g., $[-\infty, \beta_1]$, ..., $[\beta_{m-1}, \infty]$) according to the Gaussian distribution function, with each value x_i mapped to an alphabet value a_k [175]. Equation 5.4 describes the transformation from a time series X to a SAX-based time series Y using an alphabet of size m. The overall complexity of SAX is $O(n + m \cdot log(m))$.

[1] Available at MIT-BIH [227].

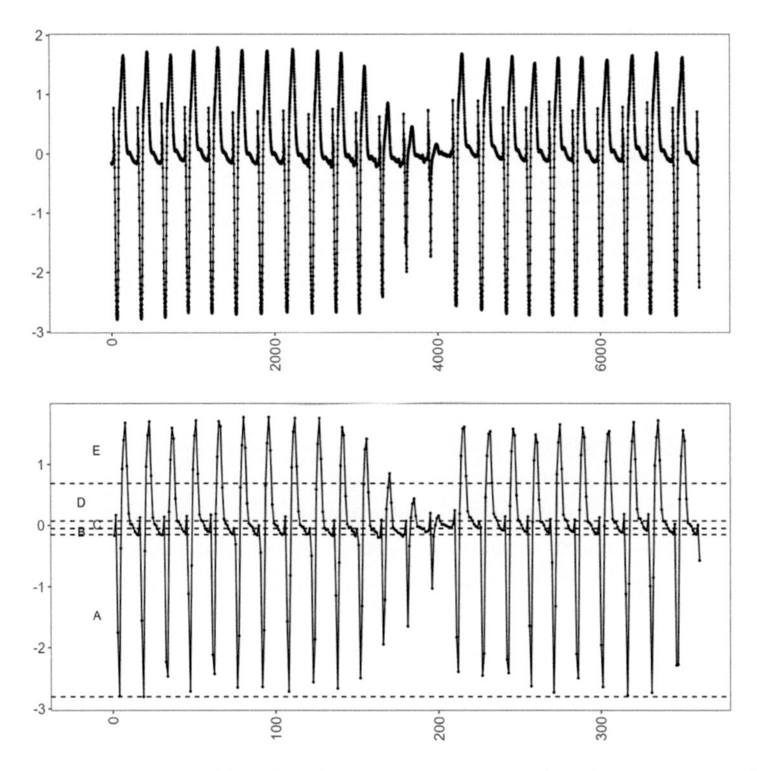

Fig. 5.2 Fraction of patient #102 heartbeat in an electrocardiogram from MIT-BIH normalized using Z-Score (**a**), PAA applied to the heartbeat example using $n' = 20$ (**b**). The figure also presents the SAX encoding thresholds for an alphabet size of five

$$y_t = \begin{cases} a_1, & x_t \in [-\infty, \beta_1], \\ a_2, & x_t \in [\beta_1, \beta_2], \\ \dots \\ a_m, & x_t \in [\beta_{n-1}, \infty] \end{cases} \tag{5.4}$$

It is common to apply SAX [175] after applying PAA as the temporal aggregation provided by PAA decreases the influence of the residual component of a time series. This enables the discovery of motifs that are more related to trend and seasonal components and, as a consequence, are more representative of a phenomenon. Furthermore, SAX indexing is usually applied on a sliding window basis, where the sliding window size is known as word size. Figure 5.2b shows intervals for converting a patient's heartbeat in an electrocardiogram into the SAX alphabet of size five. Considering a word size of four, the first sequence would be $ADAA$. This example illustrates SAX, but a larger alphabet size should be adopted for motif discovery using real-world data.

Other Indexing Techniques

Recent indexing techniques include Graph-based Indexing, Deep Learning-Based Indexing, and Hybrid-based Indexing [217]. Graph-based Indexing uses enhanced adaptive PCA combined with hierarchical navigable Small World graphs to cluster time series and build graphs for each cluster. This technique accelerates indexing, significantly reduces memory usage, and maintains high accuracy in similarity searches.

Deep Learning-Based Indexing uses deep learning architectures based on SEAnet to enhance time series indexing. SEAnet leverages neural networks to learn data series embeddings, capturing intricate patterns and relationships and facilitating faster and more accurate similarity searches than traditional methods. SEAnet is effective for high-dimensional and complex time series requiring high precision and speed [12].

Hybrid-based Indexing combines traditional indexing techniques, such as PAA and SAX, with modern techniques such as Deep Learning-Based Indexing and Graph-based Indexing. Hybrid-based Indexing leverages the strengths of each technique to balance precision and computational efficiency, applying PAA and SAX for initial data reduction and discretization, followed by deep learning embeddings or graph-based clustering, resulting in scalable time series indexing frameworks [177].

5.3 Similarity Measures

Motif identification relies on evaluating the similarity between time series subsequences to identify motif candidates, with several authors proposing efficient methods for this purpose [199, 279]. These methods are based on measuring the distance between time series and sequences, where higher similarity corresponds to lower distance. These measures are categorized into distance-based and elastic measures, as presented below.

5.3.1 Distance-Based Measures

Distance-based measures, or lock-step measures, compare corresponding points in two time series, maintaining a one-to-one correspondence throughout the series [217]. The ith point of one time series is aligned and compared with the ith point of another time series, maintaining a one-to-one correspondence throughout the entire series. Common lock-step measures include the Euclidean Distance and the Hamming Distance.

Euclidean Distance
The most widely used distance measure in time series analysis is the Euclidean Distance, which measures the distance between points from different sequences. Given two sequences q and q', both of size p, the Euclidean distance $EDist(q, q')$ is obtained by Eq. 5.5.

$$EDist(q, q') = \sqrt{\sum_{i=1}^{p}(q_i - q_i')^2} \tag{5.5}$$

Hamming Distance
The Hamming Distance between two sequences of equal length is the number of positions at which corresponding observations differ, thus measuring the minimum number of substitutions required to change one sequence into another. A Hamming Distance(q, q') is obtained by Eq. 5.6, derived from the sum of distances between each Hamming Item, as described in Eq. 5.7. The distance between two items (x and y) is zero when equal and one otherwise.

$$HDist(q, q') = \sum_{i=1}^{p} HItem(q_i, q_i') \tag{5.6}$$

$$HItem(x, y) = \begin{cases} 1, & x \neq y, \\ 0, & x = y \end{cases} \tag{5.7}$$

Other Distance-Based Measures
Other significant distance-based measures for time series analysis include Manhattan Distance, Canberra Distance, and Chebyshev Distance. Manhattan Distance, also known as L1-norm or City Block Distance, sums the absolute differences between corresponding points in two sequences, offering robustness to outliers. Canberra Distance, a weighted version of Manhattan Distance, emphasizes differences where values are small, providing sensitivity to minor changes. Chebyshev Distance, or L∞-Norm, measures the greatest difference between corresponding elements of the sequences, focusing on maximum deviation and offering robustness in scenarios where large differences are critical. These measures maintain a one-to-one correspondence between points in the series [217].

5.3.2 Elastic Measures

Elastic measures compare time series by allowing for nonlinear alignments between sequences, accommodating variations in timing and phase. Unlike distance-based measures (such as Euclidean Distance), which require a one-to-one correspondence between points [87], elastic measures handle time series with different lengths, speeds, or slight temporal distortions. This is particularly true when observing the same phenomenon multiple times where differences in elapsed time are expected [54]. This occurs in applications such as speech recognition, motion capture analysis, and financial market monitoring, as they capture underlying similarities despite temporal misalignments [217]. DTW is the most well-known elastic measure, aligning sequences by warping their time axes to minimize the distance between corresponding points. Other examples include Edit Distance with Real Penalty and Move–Split–Merge.

Dynamic Time Warping

DTW is a widely used elastic measure. It uses dynamic programming to align time-phased time series by building a matrix of quadratic distances between points. A path through the matrix that minimizes the total cumulative distance finds the best match between the series. When the goal is to compute the measure (not the warping path), then the matrix can be replaced by loading two columns simultaneously (one of each at a time series).

DTW can be computed by measuring the number of editions needed to transform one sequence into another. Given two sequences q and q', respectively, with sizes p and p', DTW measures the editing needed to transform q and q'. Editing operations for each pair of observations include: (i) replacing the next element of q with the next element of q'; (ii) elongating an element, matching the next element of q to the last element of the already-matched prefix of q' or vice versa. DTW computes the minimal transformation cost by filling the entries of a $p \times p'$ matrix ($dtw^{q,q'}$). Assuming no elongation operation and the univariate time series scenario, Eq. 5.8 describes DTW between two sequences q and q', with a further lookup on the pre-computed matrix expressed in Eq. 5.9.

$$DTWDist(q, q') = dtw^{q,q'}_{p,p'} \tag{5.8}$$

$$dtw^{q,q'}_{i,j} = min(|q_i - q'_j| + dtw^{q,q'}_{i,j-1}, |q_i - q'_j| + dtw^{q,q'}_{i-1,j}, 2|q_i - q'_j| + dtw^{q,q'}_{i-1,j-1}) \tag{5.9}$$

Figure 5.3 shows DTW between two sequences: $T_1 =< 1, 3, 2, 4, 3 >$ and $T_2 =< 1, 2, 3, 3, 4 >$. The DTW between the two sequences is 3. Figure 5.3a provides the dtw matrix, enabling follow-up on the computed distance. The solution for the computation of $dtw^{T_1,T_2}_{3,4}$ is expressed in Fig. 5.3b, which is equal to 2.

Edit Distance with Real Penalty

Edit Distance with Real Penalty is another elastic measure designed to compare time series by considering alignment and the magnitude of differences between points. Edit Distance with Real Penalty introduces a real penalty for gaps, aligning sequences with insertions and deletions. This method is effective in applications where both temporal alignment and magnitude deviations are important, such as speech and gesture recognition. By incorporating penalties for alignment errors, Edit Distance with Real Penalty ensures that comparisons reflect underlying patterns [109].

Move–Split–Merge

Move–Split–Merge addresses the shortcomings of traditional DTW by allowing flexible operations, such as moving, splitting, and merging points. This measure effectively handles

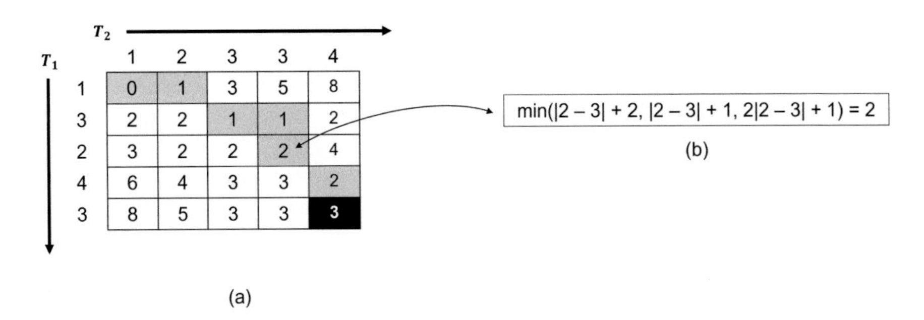

Fig. 5.3 DTW between two sequences: T_1 ($< 1, 3, 2, 4, 3 >$) and T_2 ($< 1, 2, 3, 3, 4 >$) (adapted from Buza [54])

time series with structural variations and complex temporal patterns, offering superior clustering and classification tasks. Move–Split–Merge enhances the robustness of time series analysis by accommodating various forms of temporal distortions and structural changes [136].

Other Elastic Measures

Other elastic measures handle temporal distortions and structural variations, including Longest Common Subsequence, Time Warp Edit Distance, Edit Distance on Real Sequence, Weighted Dynamic Time Warping (Weighted DTW), and Fast Dynamic Time Warping (FastDTW). The Longest Common Subsequence identifies the longest subsequences that can be matched between two different time series, useful in noisy data applications. Time Warp Edit Distance integrates both time and amplitude dimensions to compare sequences, balancing alignment flexibility and penalization of distortions. Edit Distance on Real Sequence extends traditional edit distance concepts by incorporating real-valued penalties, effective in varying data precision scenarios. Weighted DTW introduces weights to penalize certain alignments, enhancing robustness against over-compression. Finally, FastDTW offers a scalable approximation of DTW, significantly reducing computational complexity while maintaining alignment accuracy, making it suitable for large-scale time series [136, 217].

Lower Bounds

Lower bounds are fundamental in time series similarity searches, especially when using elastic distance measures. Elastic measures, such as DTW or Edit Distance with Real Penalty, allow flexible alignments and distortions but are computationally expensive. Lower bounds offer a quick, approximate distance measure that can eliminate non-promising candidates before exact computations are performed, significantly reducing computational load [218].

A lower bound provides a distance estimate that is always less than or equal to the actual elastic distance between two time series. This ensures that if the lower bound distance between a query and a candidate time series exceeds the current best match, the candidate can be discarded without further computation. This approach speeds up the overall search process by reducing exact distance calculations. Effective lower bounds are tight, meaning their values are close to the actual elastic distances, maximizing pruning power.

Various methods construct lower bounds tailored to specific elastic measures. For example, LB_Keogh, a well-known lower bound for DTW, uses envelope-based summaries of the time series, constructing upper and lower envelopes around the query time series, then computing the distance to the candidate series. This method effectively captures the global shape of the series while providing a quick distance estimate. Other lower bounds, such as LB_Improved, enhance this method by considering the projection of the candidate series onto the query envelopes, improving tightness at the cost of increased computation [168].

Effective lower bounds balance tightness and computational efficiency. While tighter lower bounds provide better pruning, they can be more complex and costly to compute. Thus, lower bounds have been proposed to strike an optimal balance, providing substantial pruning benefits without excessive overhead. The Generalized Lower Bounding framework offers a structured method for creating effective lower bounds across various elastic measures, enhancing time series similarity search efficiency and accuracy [218]. A key innovation is its ability to construct cache summaries for query and target time series, using these summaries to compute lower bounds. By leveraging these summaries, Generalized Lower Bounding approximates the distance between time series, pruning non-promising candidates before performing more expensive elastic distance computations.

5.4 Motif Discovery Methods

Motif discovery methods can be broadly organized into main groups: Brute-Force Discovery, Index-Based Discovery, Random-Projection Discovery, Early-Abandon Discovery (EAD), and Matrix Profile (MP). Methods based on indexing are also called approximate methods, while those that compute the distance between sequences to discover motifs are termed exact methods [200].

Motif discovery approaches also differ in their focus. Some methods aim to find pairs of the most similar subsequences, such as those based on the MP, which is the state of the art for pairwise-based motif discovery. Others provide a broader search, encompassing sets of sequences that are approximately similar according to an established threshold [252]. We have made representative motif discovery methods available through our publicly Harbinger R package (see Appendix A).

5.4.1 Brute-Force Discovery

The Brute-Force Discovery method is the simplest but has a high computational cost, in particular, for discovering longer sequences in large time series [200]. It is best suited for discovering shorter sequences. This method is exhaustive, performing all possible comparisons between subsequences in a time series, ensuring complete coverage and accuracy [279].

Algorithm 5.1 outlines a Brute-Force Discovery method for motif discovery. The process begins by normalizing the time series X using Z-Score. It then extracts all pairs of sequences of size q and computes their distance. If the distance is below ϵ, the occurrences are associated with the sequence p. When a sequence p has occurrences greater than or equal to σ, it is identified as a motif. Assuming n is the length of the time series, the complexity is $O(n^2)$ times the complexity of the distance function.

1: **procedure** $BruteForce(X, q, \sigma, \epsilon)$
2: $X' \leftarrow zscore(X)$
3: $motifs \leftarrow \emptyset$
4: **for** $i \leftarrow 1$ **to** $|X'| - q$ **do**
5: $p \leftarrow subseq(X', i, q)$
6: $occurences \leftarrow i$
7: **for** $j \leftarrow i + 1$ **to** $|X'| - q$ **do**
8: $p' \leftarrow subseq(X', j, q)$
9: **if** $dist(p, p') < \epsilon$ **then**
10: $occurences \leftarrow occurrences \cup j$
11: **if** $|occurrences| \geq \sigma$ **then**
12: $motifs \leftarrow motifs \cup < p, occurrences >$
13: **return** $motifs$

Algorithm 5.1: Brute-Force Discovery principle

5.4.2 Index-Based Discovery

The Index-Based Discovery method introduces a tolerance for motif discovery by indexing time series using PAA, SAX, or both, rather than computing exact distances between subsequences. These indexes incorporate a search tolerance for motifs. The Enumeration Of Motifs Through Matrix Approximation algorithm is a representative example of this category [221], with some improvements incorporating hash principles during motif discovery [305].

Algorithm 5.2 outlines an Index-Based Discovery method for motif discovery. It normalizes the time series, applies PAA and SAX, and then computes the sliding window for the indexed time series Y'. Motifs are efficiently identified by grouping and filtering sequences with occurrences greater than or equal to σ. The complexity of the algorithm is $O(n \cdot log(n))$.

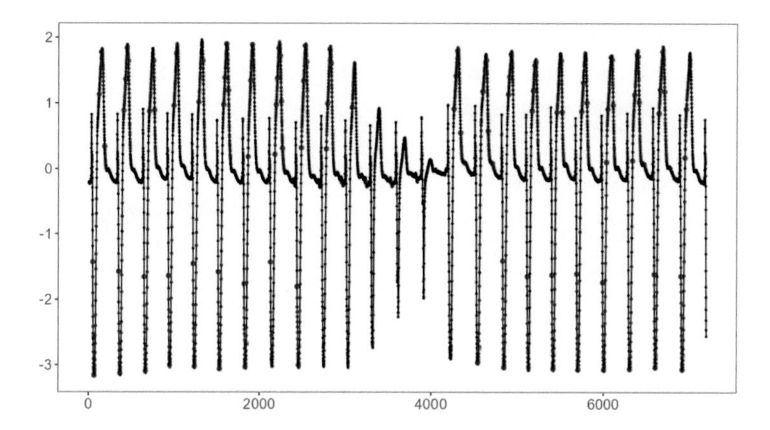

Fig. 5.4 Most similar motif with a word size of 25 detected using Index-Based Discovery for the patient #102 heartbeat in an electrocardiogram from MIT-BIH time series

1: **procedure** $IndexBased(X, a, k, q, \sigma)$
2: $X' \leftarrow zscore(X)$
3: $Y \leftarrow PAA(X', k)$
4: $Y' \leftarrow SAX(Y, a)$
5: $W \leftarrow sw(Y', q)$
6: $motifs \leftarrow group\_by\_having(W, \sigma)$
7: **return** $motifs$

Algorithm 5.2: Index-Based Discovery discovery

Figure 5.4 shows motif discovery using SAX with Index-Based Discovery for patient #102's heartbeat in an electrocardiogram from MIT-BIH. Motifs are detected in regular heartbeats, while heartbeats between 3500 and 4200 lack those regular motifs.

5.4.3 Random-Projection Discovery

The Random-Projection Discovery method addresses large time series by reducing dimensionality and randomly sampling data, optimizing execution time and computational resources in motif identification [53, 71]. In Random-Projection Discovery, similar subsequences are searched using hash-based methods for matching specific random elements of the sequence, often incorporating sliding windows and SAX. The method adds a tolerance for Hamming Distance, inspired by genetic principles related to DNA sequences with a certain degree of mutation allowed [53].

Algorithm 5.3 outlines the Random-Projection Discovery method. It assumes an indexed time series Y, checks a motif length q, and uses a maximum Hamming Distance of ext and a minimum support σ. After initial indexing, a sliding window of size $q + ext$ is created, and q columns are randomly projected from W. Motifs are then identified by filtering observations with occurrences greater than or equal to σ.

```
1: procedure RandomProjection(X, q, ext, σ)
2:     Y ← index(X)                          ▷ Assuming Z-Score and SAX
3:     W ← sw(Y, q + ext)
4:     W' ← project(W, q)
5:     motifs ← group_by_having(W', σ)
6:     return motifs
```

Algorithm 5.3: Random-Projection Discovery discovery

One interaction is insufficient for large motifs, as subsequences may not fully map as motifs, covering only a small part of the sequence. To address this, the Random-Projection Discovery process is applied multiple times, with a collision matrix built for different executions. The motif identification selects indexes with the highest values in the collision matrix, representing potential motifs, although this property is probabilistic [71]. Random-Projection Discovery can also be applied directly to numeric time series, speeding up cases where sequences are not motifs [305].

5.4.4 Early-Abandon Distance

The EAD methods optimize Brute-Force Discovery by abandoning distance computations early when the partial distance exceeds the accepted error ϵ. For example, in computing Euclidean distance (Euclidean Distance), rather than applying the square root after all components are computed, the accepted error ϵ^2 is squared, and the sum of each component is checked to see if it exceeds ϵ^2 [200]. This method has two advantages: faster distance checking and reduced search space by avoiding trivial motifs from neighboring observations. In contrast, the general complexity remains the same as Brute-Force Discovery, empirical evaluations on real data showing significant computational savings [200].

5.4.5 Matrix Profile

The MP is a distance-based data structure representing the distance of a sequence of size q to the most similar subsequence in the time series. It can use numeric data directly and employs a matrix to support subsequence comparisons, identifying similar sequences (motifs) [311]. It is called MP because a naive implementation computes all pairwise distances for all sequences in the time series, storing them in an n-by-n matrix. The minimum value in each column, excluding values near the diagonal, forms an n-length vector, the MP. This data structure also includes a MP Index, which records the location of the closest matching subsequence for each subsequence [83, 312].

MP enables all-pairs-similarity-search on two different time series. In this case, MP builds a vector containing the z-normalized Euclidean distances between each subsequence from the first series and its closest matching subsequence from the second. Formally, given two normalized time series of n observations, Y and Y', and a subsequence length q, the MP M is a matrix in $\mathbb{R}^{n-q+1}$, and a MP index is a vector I in $\mathbb{R}^{n-q+1}$. I_i contains the index of the start of the subsequence of Y' with length q that best matches $subseq(Y, i, q)$, while M_i contains the corresponding distance. In a self-join where $Y = Y'$, an additional constraint, the exclusion zone, prevents trivial matches [311].

A notable feature of the MP is its ability to detect discords, which are anomalous or unusual subsequences in the time series. By identifying the highest values in the MP, subsequences dissimilar to their nearest neighbors can be flagged as potential discords [83].

The MP was originally introduced alongside the Scalable Time Series Anytime Matrix Profile (STAMP) algorithm [311], which calculates the MP over a time series, as depicted in Algorithm 5.4. Internally, STAMP uses the MASS similarity search algorithm, which iteratively calculates the distance of each subsequence of a time series X to every subsequence of another time series X' using the FFT distance function for fast computation [312].

1: **procedure** STAMP(X, $X' = nil$, q)
2: $\quad Y' \leftarrow Y \leftarrow zscore(X)$
3: $\quad$ **if** $X' \neq nil$ **then**
4: $\quad\quad Y' \leftarrow zscore(X')$
5: $\quad M \leftarrow$ infs, $I \leftarrow$ zeros
6: $\quad$ **for** $i \leftarrow 1$ **to** $|Y'| - q$ **do**
7: $\quad\quad seq \leftarrow subseq(Y', i, q)$
8: $\quad\quad D_i \leftarrow MASS(seq, Y)$
9: $\quad\quad M_i, I_i \leftarrow eleMin(M, I, D_i, i : (i + q - 1))$
10: $\quad$ **return** $\{P, I\}$

Algorithm 5.4: STAMP Algorithm

Algorithm 5.4 works as follows. At line 5, the MP M and MP index I are initialized. From lines 6 to 9, distance profiles D are computed using each subsequence seq from Y'. The pairwise minimum for each element in D is compared with the paired element in M (e.g., $\min(D_i, M_i)$ for $i = 1$ to $|Y'|$). As the minimum pair operations are performed, I_i is updated with seq when $D_i \leq M_i$. Finally, the result M and I are returned at line 10.

When $Y = Y'$, STAMP computes the MP for general similarity join, ignoring trivial matches in D during the element-wise minimum ($eleMin$) at line 9. The overall complexity of the algorithm is $O(n^2 \, logn)$, where n is the length of the time series. Since all subsequences

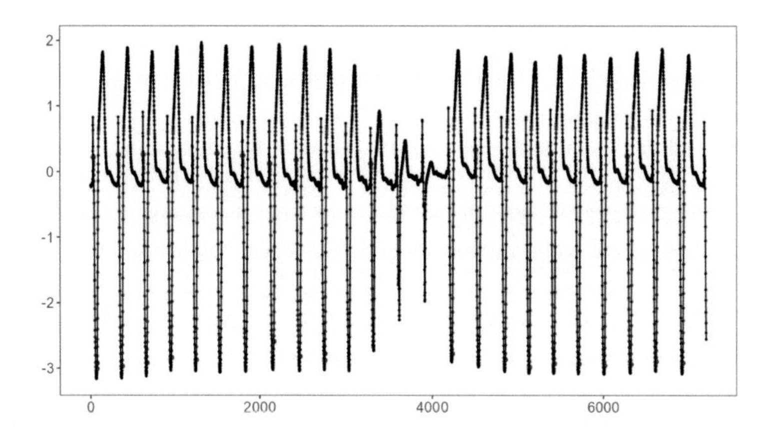

Fig. 5.5 Most similar motif with word size of 25 detected using MP for the patient #102 heartbeat in an electrocardiogram from MIT-BIH time series

are compared using the MASS algorithm, the $n \log n$ factor comes from the FFT distance subroutine invoked inside MASS. A later refinement, SCRIMP, improves the complexity, leading to a $O(n^2)$ algorithm. Although both STAMP and SCRIMP may appear slow, they are anytime algorithms, meaning they can compute a good approximation to the MP even in massive time series while converging toward the true MP.

Figure 5.5 shows motif discovery using MP with STAMP for patient #102's heartbeat in an electrocardiogram from MIT-BIH. Similar to the SAX Index-Based Discovery method, motifs are detected in regular heartbeats, while heartbeats between 3500 and 4200 lack regular motifs.

5.5 Advanced Topics

This section explores advanced topics in motif discovery, including multivariate time series, variable motif size, motif ranking, big data, and novel data structures.

5.5.1 Multivariate Motif Discovery

Various approaches address multivariate time series, which can be grouped into four main categories: (i) dimensionality reduction; (ii) combination of discovery; (iii) indexed-based discovery; (iv) exact match [179, 190, 291, 310]. These solutions can be applied separately or together.

Reduction of dimensionality. This straightforward approach applies dimensional reduction techniques, such as PCA, to transform a multivariate problem into a univariate one [274]. In some cases, due to many dimensions, time series are first projected onto a lower-dimensional subspace and then subjected to dimensionality reduction techniques for univariate motif discovery.

Combination of discovery. Univariate motif discovery methods are applied to each dimension of the multivariate time series independently, treating each as a separate univariate time series. The identified motifs are combined using logical operations (AND or OR) to form multivariate motifs [284].

Indexed-based. Indexed-based methods discretize the data and provide temporal aggregation for time series using SAX and Discrete Fourier Transform for discretization and PAA for temporal aggregation, speeding up the motif discovery process in multivariate time series [262].

Exact match. Exact match methods rely on distance measures to capture the dissimilarity between multivariate time series using a particular distance (e.g., Euclidean, DTW, correlation). These distances are then used within clustering or pattern-matching algorithms to identify motifs [16, 311].

5.5.2 Variable Motif Size

In the methods presented so far, the motif discovery methods are based on a fixed motif size (i.e., the length of the desired subsequences is fixed). However, some methods enable discovery without specifying the motif size or by establishing a range of motif sizes to search [193, 208]. For example, Tang and Liao [275] find motifs of size k (fraction of a larger motif) and then concatenate discovered motifs to search for more complex ones. Mueen [198] presents a method for discovering varying-sized motifs using the concept of cover, similar to generating candidates in sequence mining. Linardi et al. [177] introduce a framework that provides an exact motif discovery algorithm, finding all motifs and discords in a range of motif sizes.

5.5.3 Ranking Motifs and Occurrences

After motif discovery, an important task is sorting motifs according to their relevance [58]. A standard classification method is k-motif, which considers the number of occurrences in a time series. Motifs can also be sorted by relevance, with those resembling a straight line, i.e., constant observations, being lower-ranked or discarded depending on the data domain [71].

Various methods have been proposed to evaluate motif significance and relevance. One approach is information gain, which measures the likelihood of a motif's occurrence [58]. Log-odds consider the rarity of a motif by comparing its occurrences with the expected chance based on probabilistic distribution [307]. Castro and Azevedo [58] propose estimating motif frequency based on Markov Chain models, comparing actual and expected frequency using hypothesis tests.

Some motif discovery methods naturally rank motifs by similarity, such as MP, which returns pairs of motifs ranked by similarity [311]. Other methods rank motifs using different metrics, potentially influenced by the type of time series studied. Borges et al. [44] propose ranking motifs based on a balance among occurrences, motif signal entropy (for indexed motifs), and occurrence density, with closer occurrences ranked higher.

5.5.4 Big Data

Big data brings unique challenges for motif discovery, such as improving distance computation. For example, Rakthanmanon et al. [237] and Alaee et al. [17] enhance DTW computation for motif discovery. Alaee et al. [17] introduced SWAMP, based on MP. Furthermore, Mueen et al. [201] proposed a disk-aware algorithm for discovering exact motifs in large time series.

Other works focus on processing indexed time series. Castro and Azevedo [57] introduced iSAX (iSAX) for motif discovery using PAA and SAX. Some works also address the speed of processing time series collected in streaming environments. Fuchs et al. [104] introduced SwiftMotif, designed for streaming data. Novel methods to improve large time series management include the mSTAMP algorithm, which builds on MP [330].

5.5.5 Novel Data Structures and Methods

Many new data structures empower motif discovery using MP, such as context MP, Scalable Time series Ordered-search Matrix Profile (STOMP), and AAMP. Context MP establishes a searching constraint for MP, enabling the discovery of structures temporally close to each other [83].

STOMP is similar to STAMP [154] and can be seen as a highly optimized nested loop search with repeated calculation of distance profiles as the inner loop. However, STOMP performs an ordered search, reducing time complexity by $O(n \log n)$ by exploiting the locality of searches.

AAMP [12] is an efficient algorithm for computing MP with pure (non-normalized) Euclidean distance, avoiding the need for Z-Score normalization. Major improvements and novel applications of MP are detailed in Keogh [153].

5.6 Conclusion

This chapter covered motif discovery, the process of identifying recurring sequences within a time series. It provided fundamental concepts, formalizing motifs and the conditions for their occurrence. The chapter presented major preprocessing techniques for handling the computational complexity of motif discovery, including normalization using Z-Score to standardize data, PAA to reduce dimensionality by averaging segments of the time series, and SAX to convert time series into symbolic representations for easier pattern recognition. Advanced indexing techniques, such as graph-based, deep learning-based, and hybrid methods, were also presented to enhance the efficiency of motif discovery.

Similarity assessment is critical for motif discovery. Distance-based measures like Euclidean and Hamming distances offer direct comparisons, while elastic measures, such as DTW, Edit Distance with Real Penalty, and Move–Split–Merge, enable nonlinear alignments. Lower bounds provide approximate distance measures to prune non-promising candidates, improving computational efficiency.

The chapter detailed the main motif discovery methods: Brute-Force Discovery, which compares all possible subsequences; Index-Based Discovery, which enhances efficiency using indexing techniques like PAA and SAX; Random-Projection Discovery, which reduces computational load through dimensionality reduction and probabilistic methods; EAD, which optimizes brute-force methods by terminating calculations that exceed a threshold; and MP, which uses a matrix to find and rank similar subsequences.

Advanced topics were also introduced, including multivariate motif discovery, variable motif size, motif ranking, big data processing, and novel data structures and methods. These topics highlight new opportunities for innovative methods and techniques, such as addressing nonstationary time series, handling multivariate and spatial–temporal data, filtering and ranking motifs to speed up discovery, and combining index-based and distance-based methods as a promising direction for future research.

6 Online Event Detection

6.1 Online Versus Offline Scenarios

Event detection is the process of discovering events in a time series, whether by recognizing past events (offline detection) [214], detecting real-time events (online detection) [32], or predicting future events before they occur (online prediction) [325]. It is fundamental in surveillance and monitoring systems and has gained significant research attention, in particular, in applications involving large time series from critical systems [140].

Various methods have been developed for time series event detection [81], each with distinct characteristics or assumptions about time series distribution. In classification-based event detection, a time series is divided into training and test sets, with the test set containing recent data [139]. The model is built using the training set, often partitioned using time series cross-validation to optimize hyperparameters, and evaluated with the test set [125]. This traditional scenario is depicted in Fig. 6.1a.

Event detection from streaming data involves continuous observations. The data timestamp must be considered even if the entire stream is stored in batches. Training uses past data to predict more recent data, mimicking streaming scenarios. Figure 6.1b shows the i-th batch (training data) used to build a model for evaluation with the next batch $(i + 1)$ (test data). For example, this methodology is crucial when studying concept drift.

6.2 Online Detection

Online detection identifies events as observations arrive in a streaming scenario. Unlike offline event detection, where the entire time series is available, online detection requires sequential analysis of each observation as it arrives, with all processing and learning conducted online. In this case, an online or streaming time series X is composed of continuous

E. Ogasawara et al., *Event Detection in Time Series*, Synthesis Lectures on Data Management, https://doi.org/10.1007/978-3-031-75941-3_6

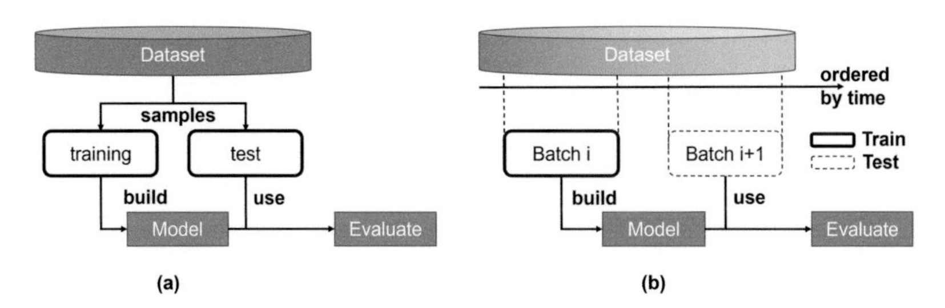

Fig. 6.1 Characterization of training and testing during classification: **a** traditional scenario; **b**: streaming scenario (adopted from Giusti et al. [110])

input: $< \ldots, x_{t-2}, x_{t-1}, x_t, \cdots >$. At each time t, a model trained on previous observations $< \ldots, x_{t-2}, x_{t-1} >$ determines whether the current behavior is unusual with the arrival of the next input x_{t+1}. The time series model might be continuously updated [9].

In online detection, time series is not split into static train/test sets, and algorithms cannot look ahead. Thus, operating in an unsupervised or semi-supervised manner is often necessary. Early detection of events in streaming applications is important, providing actionable information to prevent possible system failures. However, there is a tradeoff between fast detections and false positives, in which frequent inaccurate detections may lead to the algorithm alerts being ignored [228].

Applications requiring uninterrupted monitoring of network infrastructure events might quickly generate real-time data (fast data) [80]. In fast data, information undergoes dynamic changes within short intervals ranging from seconds to milliseconds. A substantial volume of data arrives continuously, making streamed data analytics critical for event analysis and decision-making [32].

Detecting events in fast data has significant applications across various industries, yet providing resilient solutions remains challenging [32]. Current approaches for online detection can be categorized into two groups based on their modeling methods. The first group includes static models trained on large data samples and deployed on data streams, remaining unchanged during execution. The second group involves dynamic models initialized on a data sample and then incrementally updated as new data arrives.

Accommodating large volumes of streaming data in a machine's main memory is often infeasible. Full-memory approaches, depicted in Fig. 6.2a, which consider all observations (batches) seen so far, are impractical [142]. Hence, solutions that maintain only a fraction of the data are required.

Data models can be retrained using recent data batches or trained incrementally by continuous updates. Incremental algorithms process input examples individually (or batch by batch) and update the data model after receiving each example. The model update for any new data batch is based on the previous one (Fig. 6.2b). Incremental algorithms may also access previous or most representative samples using partial (windowed) memory, as shown in Fig. 6.2c [107].

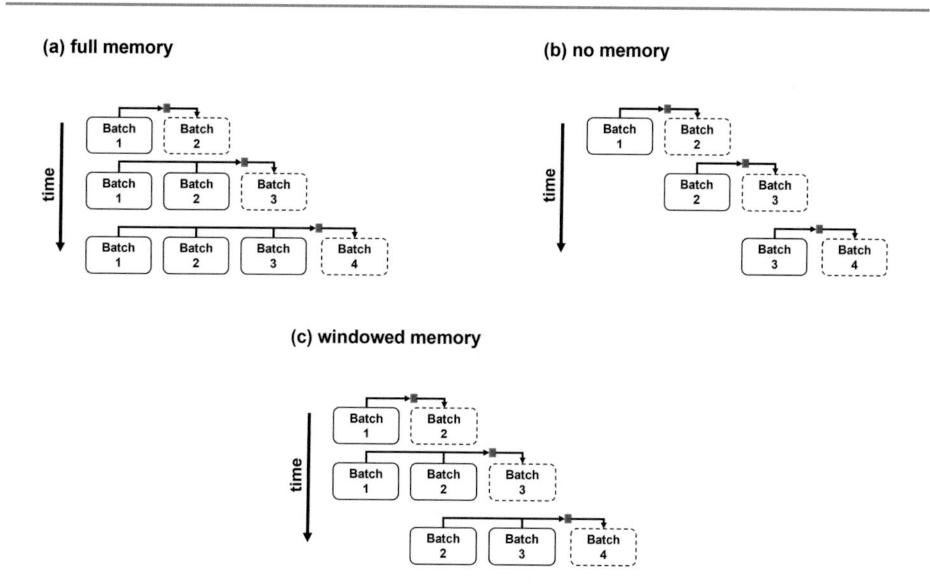

Fig. 6.2 Memory management for online event detection: full-memory, no-memory, or windowed memory. Solid batches are used for training. Dashed batches are being tested. Adapted from Iwashita and Papa [142]

Learning algorithms often need to operate in dynamic environments that can change unexpectedly. A desirable property of these algorithms is their ability to incorporate new data, leading to the stability–plasticity dilemma [89]. How can a learning system remain adaptive to significant changes while stable in response to irrelevant changes?

If the data-generating process is nonstationary, the underlying concept may change over time. Learning in nonstationary environments requires adaptive methods that monitor underlying changes and adapt models accordingly. Given that this holds for most practical applications, algorithms that effectively and efficiently learn from and adapt to evolving or drifting environments are essential. Adaptive learning algorithms extend incremental learning systems by adapting to changes in the data-generating process over time [251].

There is an intrinsic relationship between nonstationarity and concept drift [107]. When concept drift occurs, it is possible to characterize a time t when the concept of sequences before t differs from that of sequences after t. Concept drifts happen unexpectedly, but changes may occur abruptly, incrementally, or gradually, or previously seen concepts may reoccur. Models must adapt to new definitions of typical data. Adaptive learning involves real-time updating of predictive models to respond to concept drifts [107].

Adaptive learning algorithms are based on either active or passive learning. Active learning algorithms aim to detect concept drift, which is related to detecting change point events. Passive learning algorithms update the model with each new data input, regardless of drift, and may struggle to detect change points [89].

Event detection for online applications is fundamentally different from offline detection. Most existing event detection algorithms do not directly apply to streaming applications [9], creating a demand for reliable online detection methods. Over the past decade, interest in online detection applications has grown, driven by the need to analyze IoT sensor data and automated system monitoring [32]. Besides event detection in streaming data, some problems require fast and real-time reactions, known as fast event detection.

6.3 Online Prediction

Event detection and online detection identify historical or ongoing events. Online prediction, on the other hand, forecasts future events, helping mitigate potential issues through early warnings. Online prediction is similar to time series prediction but focuses on predicting whether a target event will occur within a future time window [151].

Online prediction methods use historical time series data to build a predictive model, which is then used for online predictions. These methods generally apply ML, data mining, pattern recognition, or statistical methods. Despite extensive research, online prediction methods are still in their infancy [325] and rely on time series prediction based on either classification or regression.

Online Prediction Using Time Series Classification

Online prediction using time series classification is commonly adopted [306]. Sequences and features extracted from preceding observations are analyzed to deduce temporal patterns for event anticipation. The prediction process involves applying the learned model to new time series observations to predict event labels.

Online time series prediction works are growing rapidly in the big data era. However, the need for labels (due to many unlabeled instances) and unbalanced data affect methods adhering to time series classification. Supervised learning methods require time and resources to gather labels (computationally or by experts) and instances of the rare anomalous class to support model training [196].

Online Prediction Using Time Series Regression

Combining event detection and time series regression supports online prediction. Events are detected from predicted time series observations. The historical time series $X = < x_1, \ldots, x_t >$ is used to predict future sequences $\hat{X} = < x_{t+1}, \ldots, x_{t+h} >$. Once the model is trained, prediction occurs, leading to a predicted time series $\hat{X}$. Then, events are identified using $\hat{X}$ through unsupervised or supervised event detection methods [325]. Event detection from time series regression has been studied in the context of online prediction, identifying events such as change points based on what the time series model expects [255]. Time series regression usually requires data preprocessing, as described in Chap. 2.

6.4 Advanced Topics

Only a few papers survey online detection methods. Notable examples include Zhang et al. [322], Ariyaluran Habeeb et al. [32], and Munir et al. [202]. Zhang et al. [322] investigate anomaly detection in wireless sensor networks, including limited resources and decentralized detection in such environments. Ariyaluran Habeeb et al. [32] provide a comprehensive survey of real-time big data processing methods for anomaly detection, emphasizing the scalability and demands for handling large, high-velocity, data streams. Munir et al. [202] highlight the tradeoffs among accuracy, computational efficiency, and the ability to adapt to evolving data of traditional and deep learning-based anomaly detection methods.

Additionally, Almeida et al. [21] describe general methods applied to time series analysis in big data environments, including both batch and streaming scenarios, while Fahrmann et al. [97] describe anomaly detection in smart environments, including IoT sensor data and the constraints of real-time processing in resource-limited settings.

Survey papers present analytical reviews of event monitoring and anticipation methods in time series [196]. Mehrmolaei and Keyvanpour [191] review general systems for event prediction, focusing on methods that support event prediction across application domains such as healthcare and finance. They give attention to both supervised and unsupervised learning methods. Other works, such as Salfner et al. [245], describe online failure prediction methods, including algorithms designed to predict system failures in real time through continuously monitoring system variables. Zhao [325] presents a systematic survey of event prediction methods in big data. He addresses challenges in processing large-scale time series, providing examples from disaster prediction, crime detection, and healthcare monitoring. Both Salfner et al. [245] and Zhao [325] offer taxonomies for event detection. They highlight methods for handling concept drift and adapting models in response to evolving data distributions in real-time contexts.

Online event detection based on classification requires a significant number of observations for training. Since anomalous events are rare compared to typical observations, time series are often imbalanced. Moreover, time series can change gradually (incremental drift) or abruptly (sudden drift) in nonstationary environments. In this context, unsupervised learning methods, such as time series regression, offer advantages because they do not require labeled data and can adapt to changes in the data distribution [207].

Most event detection methods have lags in detections due to the inherent delay between the analysis of previously collected observations and the arrival of new ones. These methods require inspecting recent observations before determining whether an event has occurred. Such waiting for new data introduces latency in real-time systems [23].

Offline and online event detection methods present different challenges. In offline systems, batches are processed at once, while online systems continuously process incoming streams. However, online methods can still adapt and benefit from established offline methods. Methods developed for batch processing can be modified to work with streaming through incremental learning. They can also use sliding windows to mitigate detection delays while maintaining scalability and accuracy [241].

A single time series may contain various types of events, including anomalies, change points, and motifs. Event detection methods are usually targeted to a single type of event. For example, anomaly detection methods might overlook change points, while change point detection methods might not account for anomalies or motifs. By combining specialized detection methods (anomaly detection, change point detection, and motif discovery) using hybrid methods, for example, it is possible to gain a better understanding of the different types of events occurring in a time series [23, 107, 207].

There is a growing demand for a formal definition of the different types of events. The challenge arises from the fact that while some events can be identified based on the direct analysis of the arrival observations, full and accurate detection requires additional incoming observations to confirm the event and its type. This challenge is true for change points and complex anomalies since early indicators might not be sufficient to distinguish between typical fluctuations and significant changes. The precise classification of event types can only be reliably performed after a specific lag of k observations. In such scenarios, models need to balance the urgency of detection with the accuracy of event classification, which can be difficult when dealing with concept drift or evolving time series in streaming context [97, 141].

In online multivariate time series anomaly detection, novel methods often diverge from traditional classification models due to the complexity and dynamism of multivariate data streams. Ntroumpogiannis et al. [207] present a comprehensive analysis of emerging methods, especially when handling high-dimensional real-time evolving data. They can be categorized as Distance-based, Tree-based, and Projection-based methods.

Distance-based methods are effective for real-time anomaly detection but may struggle with high-dimensional data. Iglesias Vázquez et al. [141] present the continuous monitoring of Distance-based Outliers (COD) method that identifies anomalies by continuously tracking the real-time distances between observations in a stream. This method is driven for dynamic environments where relationships between observations may change and detection of anomalies may occur without the need for retraining on static data. Similarly, Tran et al. [280] introduce the Core Point-based Outlier Detection (CPOD) method, which improves real-time outlier detection through the usage of core points. Core points summarize local regions in the data stream. CPOD reduces the computational cost associated with continuously monitoring all time series observations while maintaining high accuracy in detecting anomalies. Such focus on core points enables CPOD to adapt to changes in the data distribution.

Tree-based methods detect anomalies in large and complex datasets by partitioning the data space. Agrahari et al. [7] review the Half Space Trees (HS-Tree) method, which utilizes a hierarchical structure to detect novel patterns and anomalies in nonstationary environments. The HS-Tree method is designed for scenarios where the data distribution evolves.

It incrementally updates the model without requiring retraining on the entire dataset. The method recursively partitions the data to isolate anomalies in multidimensional time series.

Guha et al. [117] introduce the Robust Random Cut Forest (RRCF) method, which leverages random cuts in multidimensional space to identify anomalies. RRCF builds an ensemble of cut trees that provides a scalable solution for real-time anomaly detection. Each tree isolates anomalies based on their relative depth in the tree. This method detects individual and collective anomalies across large, high-dimensional datasets.

Regarding Projection-based methods, Pevný [226] introduces the Lightweight Online Detector of Anomalies (LODA). LODA is a method that utilizes random projections to reduce the dimensionality of data, enabling its usage for anomaly detection in high-dimensional data streams. LODA projects the data into multiple one-dimensional subspaces and builds histograms to model the distribution. Anomalies are detected based on data deviation from the typical observations in these subspaces. Using multiple random projections, LODA detects anomalies from different components of the time series while maintaining a low computational cost.

Manzoor et al. [188] present XStream, which is a projection-based method designed to adjust to evolving data stream features dynamically. XStream applies random projections to support real-time streaming processing. It continuously updates its model as new observations arrive. By balancing detection accuracy with computational efficiency, XStream can handle high-dimensional data streams, especially when feature space or data distribution changes over time.

Although most papers evaluate detection quality using standard classification measures such as precision, recall, and accuracy, these metrics are often inadequate for qualitatively analyzing the temporal bias of detection algorithms, i.e., their tendency to anticipate or delay event detections. Temporal bias is important since early or late detections can have implications (inability to address an issue or false alarms). Standard classification metrics do not capture this aspect well, as they focus primarily on binary outcomes (correct or incorrect classification) without considering when the event was detected relative to its actual occurrence. Moreover, standard classification measures are not fully applicable in online scenarios, where events must be detected as soon as they happen, if not before, with minimal delay. In such cases, detection timing is critical. Metrics should account for the detection accuracy and elapsed time [213].

6.5 Conclusion

This chapter presented the fundamentals of online event detection in time series. It highlighted the distinctions between online and offline detection. Unlike offline methods, where the entire dataset is available for analysis, online event detection operates in real time. Observations are processed as they arrive. This fundamental difference demands memory management, concept drift handling, and adaptable algorithms. These demands must not sacrifice computational performance.

The chapter also examined online prediction as an extension of event detection, forecasting future events based on time series classification and regression. Online prediction methods go beyond identifying past and ongoing events, offering an anticipatory capability and mitigating potential risks through early warnings.

This chapter also covered detection methods based on distances, trees, and projections. Many open research questions remain in online event detection and prediction. A key exploration area lies in efficient memory management, continuous model updates, and coping with concept drift. Balancing detection speed with accuracy is another persistent issue. While faster methods can lead to false positives, slower methods might miss critical events altogether. There are many opportunities for novel methods, including hybrid approaches that combine anomaly detection, change point detection, and motif discovery to achieve more comprehensive insights from streaming time series.

Evaluation Metrics 7

7.1 Basic Metrics

In time series analysis, an event often indicates a significant change in observations at a particular instant or over a specific interval, representing a phenomenon with defined relevance in a given domain. Event detection aims to identify such events through data analysis and is critical in surveillance and monitoring systems, particularly in applications involving time series and sensor data analysis.

Numerous methods have been proposed for event detection [325], including anomaly detection [81] and change point detection [281]. Regardless of the method used, the way events are reported is important. Assuming events are labeled, standard classification metrics such as accuracy, recall, precision, and F1 are typically employed [20]. Although accuracy is a specific metric, the term *detection accuracy* is often used to refer to a method's ability to detect events correctly.

Assessing event detection performance is essential for determining a method's suitability for a specific application. Detection performance refers to how well a method predicts the class label of data instances in event detection [276]. Table 7.1 summarizes the standard classification evaluation metrics, including accuracy, recall, specificity, precision, F1, and F_β [125].

The formulas in Table 7.1 depend on P and N, which refer to the number of positive instances (data instances of the main class of interest) and negative instances (all others). In event detection, positive instances indicate the occurrence of an event ($event = yes$), while negative instances indicate typical observations ($event = no$). Table 7.1 also shows four key metrics for comparing the class labels yielded by the detection method with the known class labels of the data instances: TP, TN, FP, and FN.

The TP metric represents the number of true positives or positive instances correctly labeled by the method. Similarly, TN refers to the number of true negatives or negative instances correctly labeled by the method. On the other hand, FP represents the number

E. Ogasawara et al., *Event Detection in Time Series*, Synthesis Lectures on Data Management, https://doi.org/10.1007/978-3-031-75941-3_7

Table 7.1 Standard event detection performance metrics [125]

Metric	Formula
Accuracy, recognition rate	$\frac{TP+TN}{P+N}$
Error rate, misclassification rate	$\frac{FP+FN}{P+N}$
Precision	$\frac{TP}{TP+FP}$
Recall, sensitivity, true positive rate	$\frac{TP}{P}$
F_1, F-score, harmonic mean of precision and recall	$\frac{2 \times precision \times recall}{precision + recall}$
Specificity, true negative rate	$\frac{TN}{N}$
Fallout, false positive rate	$\frac{FP}{N}$
F_β, where β is a non-negative real number	$\frac{(1+\beta^2) \times precision \times recall}{beta^2 \times precision + recall}$

Table 7.2 Standard confusion matrix for event detection

		Detected events		Total
		Yes	No	
Known events	Yes	TP	FN	P
	No	FP	TN	N
	Total	P'	N'	$P+N$

of false positives or negative instances incorrectly labeled as positive. Finally, FN refers to the number of false negatives, or positive instances mislabeled as negative [125, 245]. These metrics are typically summarized in a confusion matrix (see Table 7.2), which is a useful tool for analyzing how well a detection method can recognize the positive (event) class. TP and TN indicate correct labeling, while FP and FN indicate mislabels. Totals are also provided. Besides P and N, P' represents the number of data instances labeled as positive, and N' represents the number of instances labeled as negative. Given a time series X, the total number of instances is $TP + TN + FP + TN = P + N = P' + N' = |X|$.

The standard classification-based event detection performance metrics listed in Table 7.1 allow for specific performance analysis of the detection methods. The accuracy metric reflects how well the method recognizes data instances of both classes (positive and negative).

Table 7.3 Example of detections in a time series with ten observations

Time	Detector	Score	Actual
1	F	0.10	F
2	F	0.20	F
3	F	0.30	T
4	F	0.40	F
5	F	0.45	F
6	F	0.46	T
7	F	0.47	T
8	F	0.49	T
9	T	0.50	F
10	T	0.60	T

The error rate is the complement of accuracy. Sensitivity and specificity reflect how well the method can recognize positive instances ($event = yes$) and negative instances ($event = no$), respectively. Fallout is the complement of specificity. Precision and recall are widely adopted metrics: precision reflects exactness (the percentage of instances labeled as positive that are so), while recall reflects completeness (the percentage of positive instances labeled as such). A perfect precision score (1.0) indicates that every instance labeled as an event by the method is a known event, while a perfect recall score (1.0) indicates that every known event was labeled as such. Precision and recall are combined in the F_1 and F_β metrics. The F_1 metric is the harmonic mean of precision and recall, giving them equal weight. The F_β metric, on the other hand, assigns β times as much weight to recall as to precision. Both F_1 and F_β metrics are useful for evaluating the quality of event detection [276].

Consider the example of event detection shown in Table 7.3, which involves ten observations, two marked as detections. Events are labeled at times 3, 6, 7, 8, and 10. The true positives (TP), true negatives (TN), false positives (FP), and false negatives (FN) are 1, 4, 1, and 4, respectively. As a result, accuracy, precision, recall, and F_1 were 0.5, 0.5, 0.2, and 0.28, respectively.

Event detection methods frequently use the Operating Characteristic Curve (ROC) curve to visualize a method's ability to discriminate between positive instances (events) and negative instances (nonevents) by plotting the true positive rate (tpr) against the false positive rate (fpr). The ROC curve is, in particular, useful when the detector provides an event score between 0 and 1 rather than a simple binary indication of event detection. To generate the ROC curve, tpr and fpr are sorted according to the event score.

To illustrate the concept of event score, consider a regression-based anomaly detector that relies on the error distribution between a fitted regression model and actual observations. The absolute error distribution can be normalized, and a threshold can be established to define

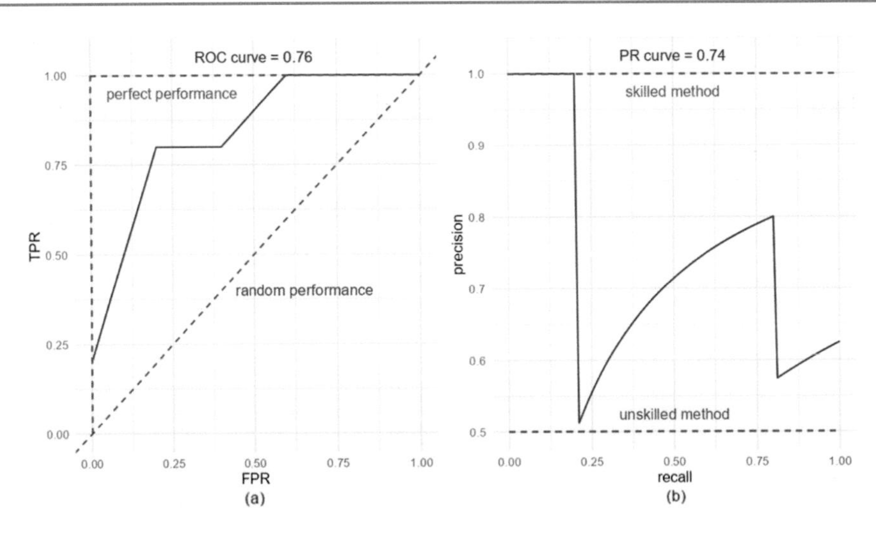

Fig. 7.1 ROC curves (**a**) and AUC-PR (**b**) for Example in Table 7.3

an anomaly. The normalized value represents the anomaly score, where values closer to 1 indicate a higher likelihood of an event, and values closer to 0 indicate a lower likelihood.

Detection methods with ROC curves closer to the upper left corner of the ROC space are the most accurate [125]. Furthermore, the accuracy of the methods can be compared using the Area Under The Curve (AUC), which summarizes the ROC curve into a single metric. The AUC can be computed using the trapezoidal rule. A perfect method achieves an AUC of 1, while a random method performs at 0.5. The ROC curve for the example in Table 7.3 is depicted in blue in Fig. 7.1a. The AUC equals 0.76, positioning it between a perfect method (in green) and a random method (in red).

$$AUC = \int_0^1 tpr(fpr)\,\mathrm{d}fpr \in [0, 1] \tag{7.1}$$

Similar to the ROC curve, the Precision–Recall Curves (AUC-PR) is defined as the area under the precision–recall curve, where recall is plotted on the x-axis and precision on the y-axis. In particular, the AUC-PR is useful in binary classification problems with imbalanced time series, which is common in event detection scenarios. When the AUC-PR approaches 0.5, it indicates poor detection capabilities, whereas values closer to 1 indicate a more accurate detector [215]. The AUC-PR is presented in Fig. 7.1b, with an AUC-PR value of 0.74, positioned between 0.5 and 1.

Another specific metric used in event detection is the Normal Precision@k, which represents the precision of the top k points with the highest anomaly scores. It can be interpreted as the precision of a method with a threshold. Considering that precision is computed as $precision = \frac{TP}{TP+FP}$, the Normal Precision@k introduces a constraint such that

$TP + FP = k$. Because of this constraint, FP is penalized, resulting in a standalone metric that does not require combination with recall for analysis [264].

7.2 Time Tolerance

The metrics presented so far focus on accurately detecting point events. However, inaccuracy in event detection does not necessarily indicate a poor result, particularly when detections occur close to the actual events. This is especially true when labels are manually marked, which may involve inherent imprecision. This section provides an example of the challenge of evaluating inaccurate event detections and highlights the need to introduce time tolerance in the evaluation process. Consider a time series X (inspired from real-world oil exploration dataset [286]) containing an event at time t, as shown in Fig. 7.2. When two detection methods, A and B, are applied to X, the user must choose the most suitable method for the underlying application. Both methods detect events close to E_1, with method A detecting an event at time 12 and method B detecting an event at time 10. However, neither method detects the event at time 11 exactly. Furthermore, only method B detects an event close to E_2 (at time 21). Based on the basic metrics, both methods would be deemed inaccurate and potentially disposable.

However, inaccuracies in detecting an event can often result from its preceding or lingering effects. For instance, consider the adoption of a new policy in a business. While a domain specialist may consider the time of policy enforcement as a company event, its effects on profit may only become detectable months later. Conversely, preparations for policy adoption may be detectable in the months leading up to its actual enforcement. Furthermore,

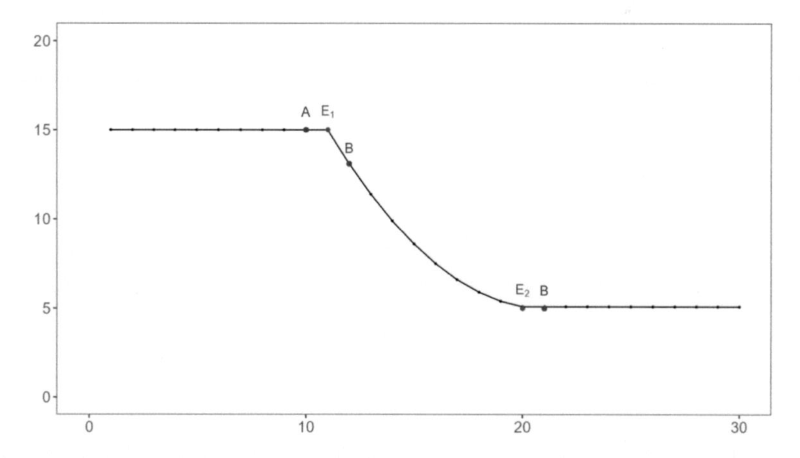

Fig. 7.2 Example of evaluating the detection of events at times 11 (E_1) and 20 (E_2). Methods A and B detect events at times 10 and 12 concerning E_1. Method B is the only method that detects an event close to E_2

when accurate detections are not feasible, detection applications often require events to be identified as soon as possible [165] or early enough to enable necessary actions to mitigate potential critical system failures or urban problems resulting from extreme weather events. In this context, the results of methods A and B would be valuable to the user. While method B seems to anticipate the event, its detection occurs afterward. On the other hand, the detection by method A is temporally closer to the event, possibly more representative of its effects. However, since this analysis uses the entire time series, it is impossible to guarantee this property. From Fig. 7.2, it can only be inferred that both methods A and B detect events at time 30, which for online event detection could be too late (see Sect. 7.4 for further discussion).

Evaluating event detection is challenging. Standard classification metrics do not account for the concept of time, which is fundamental in time series analysis, and do not reward early detection [9] or any relevant neighboring detections. In the rest of this chapter, "neighboring" or "close" detections refer to detections whose temporal distance from events falls within a desired threshold. Most research only rewards true positives (exact matches in event detection), with all other results being harshly and equally discredited.

There is a need to introduce time tolerance into the traditional concept of detection accuracy and evaluate methods while considering neighboring detection. However, state-of-the-art metrics designed for scoring anomaly detection are limited [261], often biased toward results preceding events, like those produced by method B. We have provided a step forward in addressing some of these issues by developing Soft Evaluation Event Detection (SoftED), introduced below.

SoftED

The SoftED [249] is a metric designed to assess the performance of event detection methods in time series with time tolerance. The idea behind SoftED is to introduce flexibility into traditional classification metrics (hard metrics).

Figure 7.3 illustrates the concept of SoftED, highlighting the key distinction between standard hard evaluation and the proposed soft evaluation. The standard hard evaluation involves binary outcomes, indicating whether a detection precisely matches an actual event. In contrast, the soft evaluation metrics measure the degree to which a detection is associated with a specific event.

SoftED introduces time tolerance by computing the relevance of detection to a specific event, which is determined by an event membership function $\mu_{e_j}(t)$, as defined in Eq. 7.2. The function considers an acceptable time range represented by the constant k.

$$\mu_{e_j}(t_{d_i}) = \max\left(\min\left(\frac{t_{d_i} - (t_{e_j} - k)}{k}, \frac{(t_{e_j} + k) - t_{d_i}}{k}\right), 0\right) \tag{7.2}$$

Figure 7.4a depicts the event membership function $\mu_{e_j}(t)$. For each detection d_i, the function evaluates the temporal closeness to the corresponding event e_j. The degree to which a detection d_i is relevant to an event e_j is determined by $\mu_{e_j}(t_{d_i})$, with a higher value

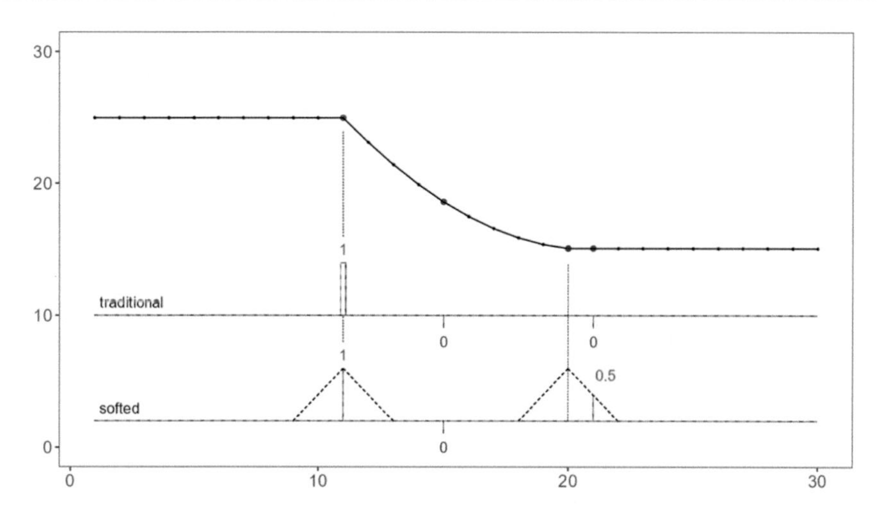

Fig. 7.3 Comparison between standard hard and soft evaluation of event detection

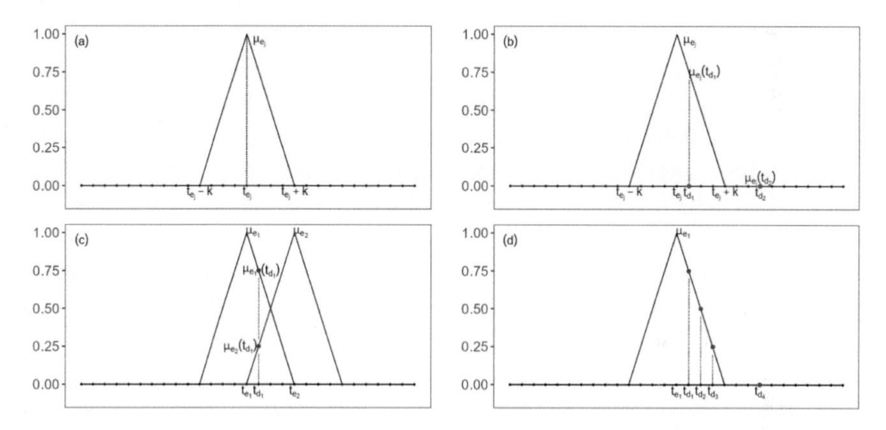

Fig. 7.4 SoftED evaluation mapping. **a** represents an event membership function $\mu_{e_j}(t)$. **b** represents $\mu_{e_j}(t)$ for detections d_1 and d_2. **c** illustrates one detection to many events, motivating the first constraint of SoftED. **d** shows many detections to a single event, motivating the second constraint of SoftED

indicating temporal closeness to a true positive (TP) regarding e_j. This concept is illustrated in Fig. 7.4b, where $\mu_{e_j}(t)$ is evaluated for two detections, d_1 and d_2, produced by a specific detection method. In this context, SoftED evaluates the extent to which a detection represents an event or, in other words, its temporal closeness to a hard TP. For instance, detection d_1 is closer to a TP. At the same time, d_2 falls outside the tolerance range and might be considered a false positive.

SoftED introduces two constraints to maintain consistency with standard hard metrics. The first constraint stipulates that a given detection d_i should have only one associated score, and the second one is that the total score associated with a specific event e_j must not exceed 1.

Table 7.4 SoftED metrics

$\mathrm{TP}_s = \sum_{j=1}^{m} \mu_{e_j}(\hat{d}_{e_j})$	$\mathrm{FN}_s = m - \mathrm{TP}_s$
$\mathrm{FP}_s = \sum_{j=1}^{m} \left(1 - \mu_{e_j}(\hat{d}_{e_j})\right)$	$\mathrm{TN}_s = (\lvert t \rvert - m) - \mathrm{FP}_s$

Consider the scenario in Fig. 7.4c, where one detection is evaluated for multiple close events. The detection d_1 is assessed for events e_1, e_2, and e_3, resulting in three distinct membership evaluations ($\mu_{e_1}(t_{d_1})$, $\mu_{e_2}(t_{d_1})$, and $\mu_{e_3}(t_{d_1})$). However, to maintain consistency with hard metrics, a given detection d_1 should not have more than one score. Otherwise, d_1 would be rewarded multiple times, potentially surpassing the score of a perfect match, which is 1.

This problem can be addressed as a bipartite graph matching problem [77], where a time series contains a set of detections $D = \{d_1, d_2, \dots d_n\}$ and a set of events $E = \{e_1, e_2, \dots, e_m\}$. The edges are the membership function for detection d_i associated with the event e_j. Representative detections ($\hat{d}_{e_j}$) are scored based on the solution of the bipartite graph matching problem, while all other detections receive a score of 0. According to Fig. 7.4d, e_1 is best represented by detection d_1, given the maximum membership evaluation of $\mu_{e_1}(t_{d_1})$. This definition ensures that the total score for detections of a specific method does not surpass the number of real events m in the time series X. Equation 7.3 maintains the reference to a perfect detection Recall score according to traditional hard metrics. Furthermore, it penalizes false positives and multiple detections for the same event e_j.

$$\sum_{j=1}^{m} \mu_{e_j}(\hat{d}_{e_j}) \leq m \tag{7.3}$$

The scores computed for each event derive soft versions of the hard metrics TP, FP, TN, and FN, as outlined in Table 7.4. The soft metrics, including TP_s, FP_s, TN_s, and FN_s, maintain the same properties and scale as traditional hard metrics. They enable the derivation of soft versions of scoring methods such as Sensitivity, Specificity, Precision, Recall, and F1. Importantly, SoftED metrics retain the same interpretability while accounting for time tolerance, which is essential in event detection applications.

The SoftED metrics are available in our Harbinger (see Appendix A), allowing the computation of both hard and soft metrics.

Other Works

The Numenta Anomaly Benchmark (NAB) provides a scoring algorithm for evaluating and comparing the efficacy of anomaly detection methods [165]. The NAB score metric is computed based on anomaly windows of observations centered around each event in a time series. Given an anomaly window, NAB uses the sigmoid scoring function to compute the weights of each anomaly detection. It rewards earlier detections within a given window and penalizes FP. NAB also allows for the definition of application profiles: standard, reward low FP, and reward low FN. Based on the window size, the standard profile gives relative weights to TP, FP, and FN.

The NAB scoring system brings challenges for its use in real-world applications. For example, the anomaly window size is automatically defined as 10% of the time series size, divided by the number of events it contains—values generally not known in advance, especially in streaming environments. Furthermore, some definitions and arbitrary constants exist in the scoring equations [261]. Finally, score values increase with the number of events and detections. Every user can tweak the weights in application profiles, making it difficult to interpret and benchmark results from other users or setups.

7.3 Interval-Based Evaluation

We now consider the scenario in which an event is interval-based. One can adopt the metrics for point events as if each observation is computed independently. However, this approach may lose important aspects of the event's semantics. The choice of evaluation metric should be guided by the nature of the time series and the specific requirements of the task at hand. Using the wrong metrics can lead to incorrect conclusions about an algorithm's performance, potentially leading to poor decisions about its use in real-world applications, such as event size and partial event detection. Specific metrics have been developed to address these issues.

For sequence events, a general evaluation prioritizes longer events, such as extended anomalies, which could indicate more severe problems that are more important to detect. This is why long anomalies contribute more to the final score than shorter ones in many metrics. Furthermore, certain metrics emphasize partial detection. While many conventional metrics aim to accurately predict each time point to correctly identify the location and duration of an event (known as *covering*), it is often more important, or at least sufficient, to detect any portion of the event (referred to as *partial detection*). It is essential to note that a human reviewing the detections may not identify an event if it is subtle and lasts much longer than the detection itself, making the detection's location and duration significant [264]. This applies to both extended anomalies and short change points.

Most metrics found in the literature are designed to evaluate point anomaly detection. However, many real-world events extend over an interval (range-based). Motivated by this, some works extend the well-known scores (precision, recall, and F_1) to address interval-based anomalies [215, 276], which, in most cases, can be generalized for interval-based events. These include Point-Adjusted f-score, Delay thresholded Point-Adjusted f-score, Segment-wise f-score, and Range-based f-score, which are explained below.

The Point-Adjusted f-score assumes that if a human can examine and identify the entire event if a single point within an event segment is accurately detected, the entire contiguous segment is marked as an event in the prediction before calculating the instant score. The Delay thresholded Point-Adjusted f-score is an adaptation of the Point-Adjusted f-score metric, where a labeled event is only considered detected if an event is detected within the first k time steps of the event. Otherwise, all points in the event are marked as false negatives, even those detected as events.

The Segment-wise f-score treats a contiguous segment of an event as a single event. One true positive is recorded for each true event segment with at least one detected event, one false negative for the remaining true event segments, and one false positive for any detected interval-based event without any true event observations. A limitation of this metric is that extending the length of a detected event does not worsen the scores.

The Range-based f-score computes recall for each labeled event using a formula that scores how well the labeled event is detected. The score is the average ratio of observations detected across all labeled events. Similarly, local precision is computed for each detected event by scoring how well a detected event corresponds to the labels. These scores are then averaged across all detected events. This metric introduces the concept of partial coverage.

7.4 Online Detection Evaluation

In many cases, especially in studies and applications where real-time systems require immediate attention to detected events, detecting a potential event early is crucial. Late detection is ineffective in such scenarios since it is too late to address the problem. Therefore, metrics should prioritize early event detection over later detections when selecting algorithms for identifying anomalies quickly [264].

This section clarifies the notion of detecting events in the early stages. Consider detection methods A and B applied to an online time series X. This example examines how these methods detect events as each new observation is made. Figure 7.5 illustrates offline and online evaluations. At time 11, shown in Fig. 7.5a, the event E_1 appears in the time series. However, neither method A nor method B detects it.

In Fig. 7.5b, at time 13, Method B detects an event close to E_1 (with a difference of one observation). Furthermore, it takes two observations to detect it. At this point, Method A still does not detect the event.

By time 15, as shown in Fig. 7.5c, Method A finally detects an event close to E_1, with a difference of two observations. This detection is not as close as Method B's detection and takes four observations to detect. It is important to note that detecting an event before it occurs does not necessarily mean anticipating detection but is related to the detection's timing. Anticipation involves detecting the event before it occurs, which would be related to event prediction.

Finally, at time 28, both methods fail to detect event E_2. Both methods make detections close to event E_1 (with a difference of one observation). However, Method A shows inconsistent detections. At time 15, it detects the event at time 9 concerning E_1, but at time 28, it

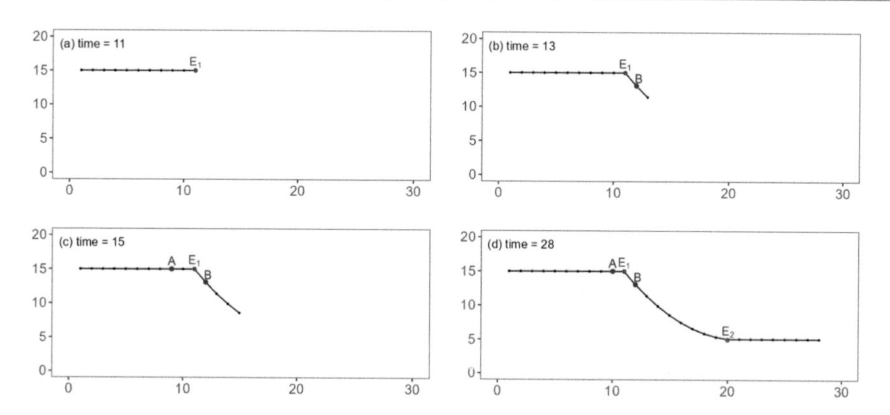

Fig. 7.5 Evaluation of two event detection methods in an online time series: time 11 (**a**), time 13 (**b**), time 15 (**c**), and time 28 (**d**)

detects the event at time 10 concerning E_1. During online event detection, actual detections may change over time. We have proposed two novel metrics to address this issue: Detection Probability and Detection Lag [174].

To explain these concepts, assume that a time series arrives in batches (b_j), each containing s observations. In a fully online context, s can be as small as one observation. Furthermore, in a streaming scenario, it is possible to adopt either a fixed memory approach, where the system preserves m batches in memory, or a full-memory approach, where all n time series observations are preserved. Note that n increases as each new batch b_j arrives.

The Detection Probability metric assesses the likelihood of detecting an event in a time series by evaluating each observation x_i across batches during processing. If x_i appears in batch b_j, it is available for detection $bf(x_i)$ times, defined as either m (partial memory) or $\lceil \frac{n}{s} \rceil$ (full memory). The set B_i contains the most recent batches where x_i is present, allowing a detector to conduct multiple evaluations to determine if x_i is an event. The detection frequency $df(x_i)$, representing the number of times x_i is marked as an event, ranges from 0 to $bf(x_i)$. The Detection Probability metric is computed as the ratio of $df(x_i)$ to $bf(x_i)$. For instance, if x_3 and x_5 are present in 7 batches, with $df(x_3) = 2$ and $df(x_5) = 6$, their respective DPs are 0.29 and 0.86.

The Detection Lag metric assesses the early detection capability of methods by measuring the lag between the first reading of an observation and its detection as an event. For an observation x_i in a time series X, sb_i represents the start batch, and fdb_i is the first detection batch. The lag, Lag_i^s, is computed as $fdb_i - sb_i$. For example, if x_9 starts appearing in batch 3 and is detected in batch 4, then Lag_9^3 equals 1. Different batch sizes can be compared using an alternative Detection Lag value in the number of observations, computed by $(fdb_i - sb_i + 1) \cdot s$.

7.5 Benchmarks

Just as important as adopting the right metrics is using appropriate benchmark time series to evaluate the performance of available and novel methods [216]. Some issues regarding the quality of available benchmark time series can provide a false sense of progress in the field. Many commonly used benchmark time series suffer from flaws such as trivial anomalies, unrealistic anomaly density, mislabeled ground truth, and a high ratio of anomalies at the end of the time series. To address these issues, the UCR Anomaly Archive provides a repository of curated time series for event detection evaluation [299]. Other repositories uses real-world labeled time series [286].

Metrics such as Relative Contrast and Normalized Clustered Events have been developed to support measuring the quality of the time series used for benchmarking [215]. The Relative Contrast computes the ratio of the expectation of the mean distance between events, as closer events tend to be more challenging to detect. The Normalized Clustered Events establishes the distance ratio between typical sequences and event subsequences. A larger NC indicates that abnormal points are closer together, increasing the difficulty of anomaly detection.

7.6 Advanced Topics

This section highlights additional works related to analyzing and comparing event detection performance. Sorbo and Ruocco [264] present a comprehensive survey of evaluation metrics for anomaly detection. Recent works have also focused on developing benchmarks for evaluating anomaly detection methods. Jacob et al. [143] provide a comprehensive benchmark for explainable anomaly detection over high-dimensional time series. Conversely, a benchmark to assess the benefits and limitations of anomaly detectors and detection metrics was provided by Boniol et al. [41].

Standard classification metrics are generally used to evaluate methods' ability to distinguish typical from anomalous observations. Aminikhanghahi and Cook [23] review traditional metrics for change point detection, such as sensitivity, G-mean, F-metric, ROC, PR-Curve, and MSE. Detection evaluation metrics have also been explored in the areas of sequence data anomaly detection [62], time series mining and representation [87], and sensor-based human activity learning [293].

Metrics in the literature are mainly designed to evaluate the detection of point anomalies. However, many real-world events extend over intervals (range-based). Motivated by this, Tatbul et al. [276] and Paparrizos et al. [215] extend the well-known Precision and Recall metrics and the AUC-based metrics to measure the accuracy of detection algorithms for range-based anomalies. Other recent metrics developed for detecting range-based time series anomalies are also included in the benchmark by Boniol et al. [41]. Furthermore, Wenig et al. [296] published a benchmarking toolkit for algorithms designed to detect anomalous subsequences in time series [40, 43].

7.7 Conclusion

This chapter covered event detection methods, focusing on point and interval events. Traditional metrics such as accuracy, recall, precision, and F_1 were presented, highlighting their limitations for event detection. The chapter also covered time tolerance, showing the need for metrics that reward near-miss detections, not just exact matches.

The chapter presented SoftED, a metric that provides time tolerance and scores detections based on their temporal closeness to actual events. It also introduced interval-based metrics, including the Point-Adjusted f-score, the Delay thresholded Point-Adjusted f-score, the Segment-wise f-score, and the Range-based f-score. These metrics also account for partial detection.

Early event detection is important in real-time systems. The chapter introduced new metrics like Detection Probability and Detection Lag to evaluate the performance of online event detection methods, focusing on the timeliness and consistency of detections. Recent works (including benchmarks and surveys) were also covered. Besides, combining time tolerance, online event detection, and interval-based metrics is a roadmap for novel studies.

Conclusion 8

8.1 Overall Research Area

Time series event detection is a broad and active research area that includes anomaly detection, change point detection, and motif discovery. Numerous research papers have been published in this area. We conducted a systematic map study of the area to better understand the scientific production in event detection. Four major queries were executed on Scopus, including journal papers, conference papers, reviews, and books. These queries, executed on November 29th, 2023, focus on time series by including the "time series" constraint to target concepts related to event detection within this context.

The first query focuses on the general concept of event detection, including both event detection and prediction, i.e., ("event detection" or "event prediction") and "time series". This query returns 693 documents, summarized in Table 8.1.

The second query focuses on anomalies, i.e., ("anomaly") and "time series", which is the most studied area. It returns 9635 documents, comprising 6679 papers, 2825 conference papers, 110 reviews, and 21 books. The third query focuses on change points in time series, including change points and concept drift, i.e., ("change point" or "concept drift"), and "time series". It returns 2310 documents, including 1719 papers, 555 conference papers, 29 reviews, and seven books. The final query focuses on motifs, i.e., "motif" and "time series", returning 693 documents, 388 journal papers, 296 conference papers, and nine reviews.

Figure 8.1 illustrates the evolution of this research area, with the y-axis presenting the number of documents (on a logarithmic scale) indexed in Scopus. Since anomaly detection is the most studied subarea, it is unsurprising that the first paper dates back to before 1960. After 1980, there was a significant increase in published documents related to change points, with general event detection beginning around 2000 and motif studies commencing in 2003 at a steady rhythm. The similar exponential growth of these areas is evident in the log scale on the y-axis.

E. Ogasawara et al., *Event Detection in Time Series*, Synthesis Lectures on Data Management, https://doi.org/10.1007/978-3-031-75941-3_8

Table 8.1 Publications according to event detection areas

Type	Event detection	Anomaly	Change point	Motif
Article	364	6679	1719	388
Conference paper	324	2825	555	296
Review	5	110	29	9
Book	–	21	7	–

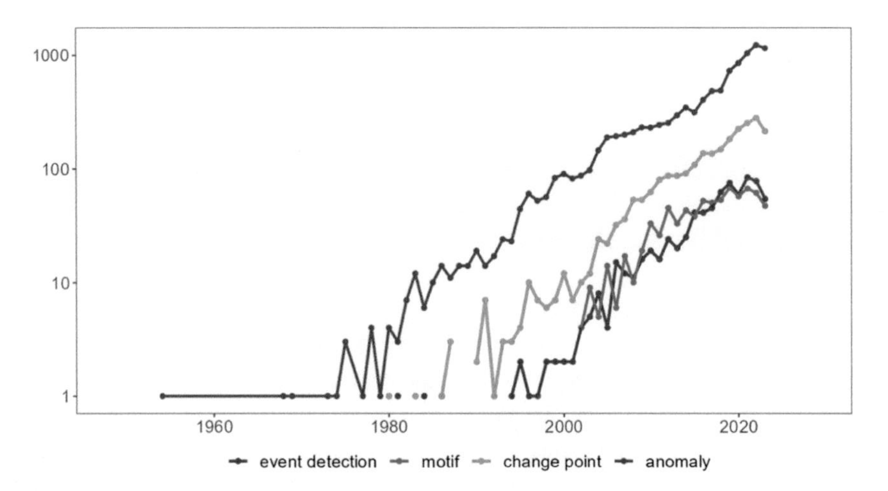

Fig. 8.1 Number of publications (in logarithm scale) in event detection according to its main areas

Table 8.2 Intersection between areas

Areas	Amount
Event detection–Anomaly	89
Event detection–Change point	13
Event detection–Motif	4
Anomaly–Change point	215
Anomaly–Motif	65
Change point–Motif	7

Since some document overlap is expected among these areas, Table 8.2 clarifies this aspect. The first interesting observation is the connection between general event detection and its subareas (anomaly, change point, and motif). The major intersection of event detection is related to anomalies (89 out of 693 documents), which includes more than 12% of documents. Both change points and motifs are also related to event detection.

The second analysis explores how anomalies, change points, and motifs relate. This investigation is interesting since some change point methods can also find anomalies. As expected, the major intersections are related to anomalies with change points. Other relationships exist between anomalies and motifs (65 documents) and change points and motifs (7 documents).

To provide a qualitative perspective of the subjects presented in the documents, Fig. 8.2 depicts a word cloud for the 100 most frequent words (after removing English stop words and applying basic text mining preprocessing) in the queried documents. Since the query string always includes "time series", these words were also removed from the analysis. There are four word clouds: Fig. 8.2a relates to general event detection (including event prediction), while Fig. 8.2b–d is related to anomalies, change points (with concept drift), and motifs, respectively.

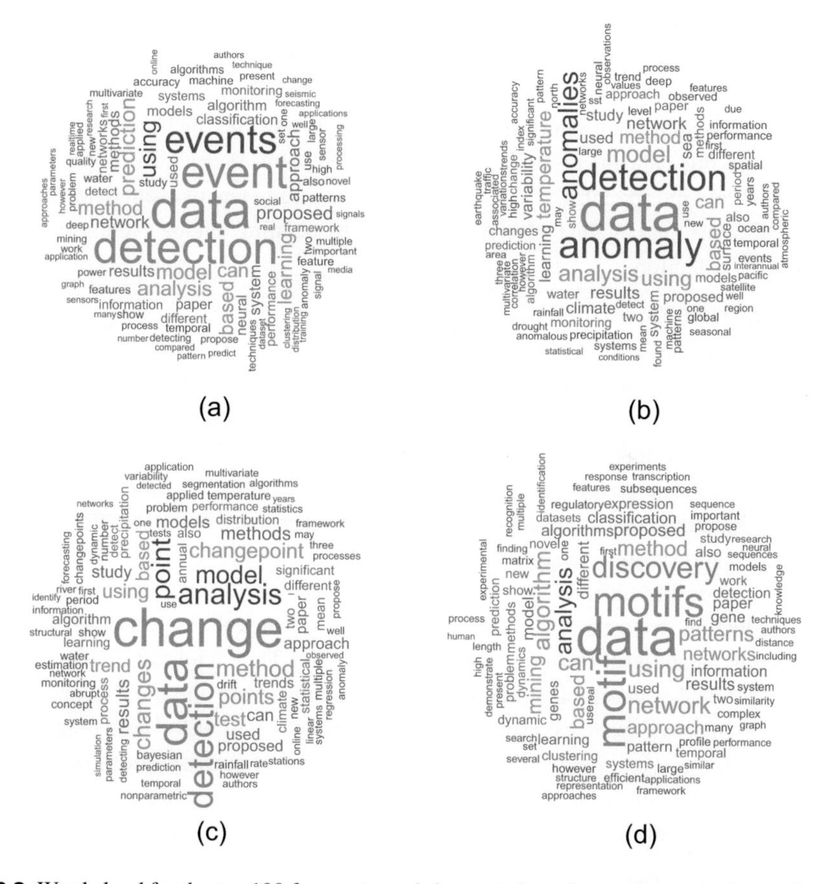

Fig. 8.2 Word cloud for the top 100 frequent words in event detection: major area of event detection (**a**), anomaly (**b**), change point (**c**), and motif (**d**)

One important difference between general event detection and the other subareas is the presence of fewer terms for application areas. General event detection has more terms related to data mining, ML, and systems. However, many climate-related terms in the area of anomalies indicate that many published papers are domain-driven. Change point papers usually contain many statistical terms, as statistical theories support many of these methods.

Motif documents have a different general vocabulary, with many terms associated with patterns, subsequences, and specific data mining functions. Notably, in Fig. 8.2a–c, the word "detection" is prominent. In contrast, in Fig. 8.2d, the related concept of "discovery" stands out.

8.2 Open Challenges in Event Detection

Throughout the book, several research challenges have been identified in event detection. While some have been partially addressed, many opportunities exist for further exploration and innovation. This section summarizes the main open challenges we believe will be targeted by the research community in the next years. It includes multivariate time series, scalability and big data integration, online event detection, metrics for event detection, explainability and interpretability in event detection, unified view of event detection, transfer learning, robustness against adversarial attacks, associating among events, and energy-efficient event detection.

Multivariate Time Series

Multivariate time series tend to become increasingly common, as many modern data sources like IoT devices and sensors often capture several variables over time. Although multivariate methods, such as vector autoregressive models and extensions of anomaly detection methods, have been developed, they still face significant limitations when dealing with high-dimensional time series. The curse of dimensionality poses a major challenge in this context, and building effective event detection methods is needed to capture the interdependencies and relationships among these variables.

Data preprocessing techniques, including transformation, decomposition, and dimensionality reduction, have been used to support detection models by reducing complexity and extracting meaningful features from the time series. They tend to be more and more applied. Additionally, ensemble models that combine univariate and multivariate methods have space to improve accuracy, especially when combined with novel data preprocessing. In this context, there is a demand for the interpretability of such solutions, especially in domains like healthcare, where understanding how events are detected is as important as providing the right detection.

Scalability and Big Data Integration

As the volume of data generated from sensors, social media, and other sources grows, scalability is becoming an important challenge in event detection. Traditional methods struggle to process large-scale and high-dimensional time series. There is an open room for building novel methods integrating big data frameworks like Apache Spark, Apache Kafka, and Apache Flink. The goal is to provide speedup in event detection solutions, especially in real-time and streaming time series.

However, many problems remain regarding optimizing these technologies for building event detection workflows. One major issue is ensuring that distributed computing systems can process large time series in real time, especially when events need to be detected as they occur. Developing parallel and distributed systems solutions for time series event detection in large-scale applications is a research roadmap.

Online Event Detection

The challenge of detecting events as soon as possible in online (real-time) scenarios has increased with the rise of data stream systems. Online event detection methods must process time series on the fly without processing the entire time series. This constraint makes detecting real-time events, such as change points and motifs, difficult. While detecting anomalies might seem simpler, identifying them near change points or in concept drift might lead to misclassifications.

Lightweight and efficient algorithms that adapt to dynamic environments are relevant in many applications, including fraud detection, industrial monitoring, and cybersecurity applications. It is common to use adaptive windowing or sliding windows as an enabling technique to support detector development. Combining these solutions while handling concept drift is very important. The goal is to support continuous model updates without too much retraining.

Metrics for Event Detection

Evaluating the performance of event detection methods is critical and challenging, especially when considering interval events. Providing time tolerance is important for applications like network traffic monitoring or fault detection in industrial systems, where detecting the interval-based event is needed for timely interventions. Some metrics provide time tolerance for point events, but time-tolerant metrics for interval events are still lacking. Novel metrics should provide a uniform basis to support interval and point events. They should also be adequate for measuring online and streaming time series detection performance. Such metrics must evolve dynamically in online detection scenarios as more data becomes available.

Explainability and Interpretability in Event Detection

With the growing use of machine learning in event detection, a significant challenge is supporting the explainability and interpretability of detectors. Detectors based on deep learning, such as LSTM and CNN, often operate as "black boxes," making it difficult to explain detected events. Understanding why detections occur is sensitive in fields like healthcare, finance, and cybersecurity.

Developing interpretable models or enhancing the transparency of existing models is important for increasing trust and adoption in real-world scenarios. Methods such as SHapley Additive exPlanations (SHAP) and Local Interpretable Model-agnostic Explanations (LIME) can inspire solutions to provide explainability to detectors. They can be supported by interactive visualization tools that can help users better understand detections and the reasons behind model decisions.

Unified View of Event Detection

Event detection is partitioned into distinct types: anomaly detection, change point detection, and motif discovery. Focusing on only one type of event may limit our understanding of real-world time series. They often exhibit these three types of events simultaneously. For instance, anomalies and change points are often related, as change points frequently manifest anomalous observations. A unified view integrating methods from different types of events could provide a more holistic view of time series dynamics. Besides, combining methods from different types could enhance the detection accuracy and robustness of a broader event detector.

Transfer Learning and Cross-Domain Adaptability

Another emerging challenge is the application of transfer learning and domain adaptation to event detection. Most event detection models are trained on specific datasets, which limits their applicability to new domains. In this case, transfer learning could be applied to adjust models for related time series, especially when labels are scarce or unavailable. Adapting these methods to multivariate time series is still an underexplored area.

The rise of LLMs has recently opened new opportunities for empowering transfer learning in time series event detection. LLMs discretize time series and treat them as sequences of tokens analogous to sentences. These models can be pre-trained on vast datasets and fine-tuned for specific tasks with minimal labeled data.

Robustness Against Adversarial Attacks

In fields such as cybersecurity and finance, time series event detectors might be exposed to adversarial attacks. In this scenario, small, imperceptible changes to input data can deceive the model into making incorrect predictions. Ensuring that event detectors (especially machine learning models) are robust against such attacks is a growing challenge.

Associating Among Events

Associating events with one another is similar to sequence mining, a data mining task that discovers patterns in ordered data. Applying this principle to time series event detection could help identify causal relationships between events. For instance, detecting a series of small anomalies could indicate the likelihood of a significant future change point in a financial market. However, mining these relationships, especially in high-dimensional, multivariate time series, is computationally intensive and demands scalable solutions, as previously mentioned.

Energy-Efficient Event Detection

Energy efficiency becomes needed in event detection in resource-constrained environments, such as IoT devices or edge computing. Many existing algorithms require substantial computational resources, which may not be feasible in low-power devices. Developing energy-efficient methods that can operate in such constraints is important for deploying solutions in applications in smart cities, wearables, and sensor networks.

Harbinger: An Unified Event Detection Framework

Considering all the challenges presented, testing state-of-the-art methods to solve event detection problems is essential. A solution that evaluates numerous alternative methods and seamlessly enables the development of novel ones is necessary to advance the area. In this spirit, we developed the Harbinger software [248], which is available as an R Package.[1]

Harbinger is a comprehensive framework designed to detect events within a time series, offering a unified environment for identifying time series anomalies, change points, and motifs. It includes a diverse array of methods for event detection, together with functions for visualizing and assessing the detected events, making it an essential tool for time series analysis. It is organized into anomaly detection, change point detection, and motif discovery.

To install Harbinger, just use the following command:

```
1  install.packages("harbinger")
```

and load the package with:

```
1  library(harbinger)
```

The library includes a sample time series to help explore its usage. For example, the code below plots the global temperature yearly time series (YGT) provided in Chap. 1:

```
1  data(examples_harbinger)
2  yts <- examples_harbinger$global_temperature_yearly$serie
3  ts.plot(yts)
```

Anomaly Detection

Harbinger uses various methods for anomaly detection, including ML models such as Conv1D, ELM, MLP, LSTM, Random Forest Regression, and SVM. It also includes classification models like Decision Tree, KNN, Naive Bayes, and Random Forest and cluster-

[1] https://cefet-rj-dal.github.io/harbinger.

E. Ogasawara et al., *Event Detection in Time Series*, Synthesis Lectures on Data Management, https://doi.org/10.1007/978-3-031-75941-3

ing methods such as K-Means, DBScan, and DTW. Additionally, statistical methods like ARIMA, FBIAD, and GARCH are incorporated.

For a full list of methods and examples, refer to the package documentation.[2] The following example demonstrates anomaly detection using ARIMA:

```
1 model <- hanr_arima()
2 model <- fit(model, yts)
3 detection <- detect(model, yts)
4 grf <- har_plot(model, yts, detection)
5 plot(grf)
```

In this example, an ARIMA detector is established, fitted to the time series, and used to detect anomalies. The results are then plotted.

Change Point

For change point detection, Harbinger employs methods such as AMOC, PELT, Chow Test, GFT, SCP, and CF. The following example uses the Chow Test test method:

```
1 model <- hcp_chow()
2 model <- fit(model, yts)
3 detection <- detect(model, yts)
4 grf <- har_plot(model, yts, detection)
5 plot(grf)
```

In this example, a Chow test detector is established, fitted to the time series, and used to detect change points. The results are then plotted.

Motif Discovery

Harbinger uses methods such as Index-Based Discovery and Matrix Profile for motif discovery, effectively identifying recurring patterns in time series. These methods are useful in various applications, including finance, healthcare, and IoT data analysis. Harbinger also supports multivariate time series.

For example, the following code demonstrates motif discovery using Index-Based Discovery on patient #102's heartbeat data from the MIT-BIH time series:

```
1 library(harbinger)
2 data(examples_motifs)
3 data <- examples_motifs$mitdb102
4 ts.plot(data$serie)
```

[2] https://cran.r-project.org/web/packages/harbinger/readme/README.html.

The following code performs motif discovery:

```
1  model <- hmo_sax(26, 25)
2  model <- fit(model, data$serie)
3  detection <- detect(model, data$serie)
4  print(detection[detection$event,])
5  grf <- har_plot(model, data$serie, detection)
6  plot(grf)
```

Evaluation

Harbinger offers evaluation metrics for detections, including traditional and soft computing (SoftED). The following example demonstrates the evaluation using a labeled synthetic time series and a ML regression model from the DAL Toolbox:

```
1  data(examples_anomalies)
2  ts <- examples_anomalies$simple
3
4  set.seed(1)
5  library(daltoolbox)
6  mlpmodel <- ts_mlp(ts_norm_gminmax(),
7       input_size=5, size=3, decay=0)
8  model <- hanr_ml(mlpmodel)
9  model <- fit(model, ts$serie)
10 detection <- detect(model, ts$serie)
11
12 evaluation <- evaluate(model, detection$event,
13      ts$event)
14 print(evaluation$confMatrix)
```

```
          event
detection TRUE  FALSE
TRUE      0     6
FALSE     1     94
```

The example above shows the evaluation results. The following code uses SoftED with a time tolerance of five observations:

```
1 result <- evaluate(model, detection$event,
2     ts$event, evaluation=har_eval_soft(sw_size=5))
3 print(result$confMatrix)
```

```
          event
detection TRUE  FALSE
TRUE      0.8   5.2
FALSE     0.2   94.8
```

Updates and More Examples

New methods are constantly being introduced in Harbinger. As of mid-2024, there are more than 50 methods for anomaly detection, change point detection, and motif discovery. Additionally, all examples presented in this book using Harbinger are available on GitHub (https://github.com/eogasawara/TSED). This repository provides practical examples and use cases, helping users quickly get started with the framework and apply it to their data.

References

1. Abdel Wahab, O.: Intrusion Detection in the IoT Under Data and Concept Drifts: Online Deep Learning Approach. IEEE Internet of Things Journal **9**(20), 19,706 – 19,716 (2022). https://doi.org/10.1109/JIOT.2022.3167005
2. Abdi, H., Williams, L.J.: Principal component analysis. Wiley Interdisciplinary Reviews: Computational Statistics **2**(4), 433 – 459 (2010). https://doi.org/10.1002/wics.101
3. Abhaya, A., Patra, B.K.: An efficient method for autoencoder based outlier detection. Expert Systems with Applications **213** (2023). https://doi.org/10.1016/j.eswa.2022.118904
4. Afzalan, M., Jazizadeh, F., Wang, J.: Self-configuring event detection in electricity monitoring for human-building interaction. Energy and Buildings **187**, 95 – 109 (2019). https://doi.org/10.1016/j.enbuild.2019.01.036
5. Aggarwal, C.C.: Data Mining: The Textbook. Springer (2015)
6. Aggarwal, C.C.: Outlier Analysis. Springer International Publishing (2016)
7. Agrahari, S., Srivastava, S., Singh, A.: Review on novelty detection in the non-stationary environment. Knowledge and Information Systems **66**(3), 1549–1574 (2024). https://doi.org/10.1007/s10115-023-02018-x
8. Aguilar, D.L., Medina-Perez, M.A., Loyola-Gonzalez, O., Choo, K.K.R., Bucheli-Susarrey, E.: Towards an Interpretable Autoencoder: A Decision-Tree-Based Autoencoder and its Application in Anomaly Detection. IEEE Transactions on Dependable and Secure Computing **20**(2), 1048 – 1059 (2023). https://doi.org/10.1109/TDSC.2022.3148331
9. Ahmad, S., Lavin, A., Purdy, S., Agha, Z.: Unsupervised real-time anomaly detection for streaming data. Neurocomputing **262**, 134 – 147 (2017). https://doi.org/10.1016/j.neucom.2017.04.070
10. Ahmed, D.M., Hassan, M.M., Mstafa, R.J.: A Review on Deep Sequential Models for Forecasting Time Series Data. Applied Computational Intelligence and Soft Computing **2022**(1), 6596,397 (2022). https://doi.org/10.1155/2022/6596397
11. Ahmed, M., Mahmood, A.N., Islam, M.R.: A survey of anomaly detection techniques in financial domain. Future Generation Computer Systems **55**, 278 – 288 (2016). https://doi.org/10.1016/j.future.2015.01.001

E. Ogasawara et al., *Event Detection in Time Series*, Synthesis Lectures on Data Management, https://doi.org/10.1007/978-3-031-75941-3

12. Akbarinia, R., Cloez, B.: Efficient Matrix Profile Computation Using Different Distance Functions (2019). https://doi.org/10.48550/arXiv.1901.05708. http://arxiv.org/abs/1901.05708
13. Akoglu, L., Tong, H., Koutra, D.: Graph based anomaly detection and description: A survey. Data Mining and Knowledge Discovery **29**(3), 626 – 688 (2015). https://doi.org/10.1007/s10618-014-0365-y
14. Akpinar, M., Yumusak, N.: Year ahead demand forecast of city natural gas using seasonal time series methods. Energies **9**(9) (2016). https://doi.org/10.3390/en9090727
15. Al Shalabi, L., Shaaban, Z.: Normalization as a Preprocessing Engine for Data Mining and the Approach of Preference Matrix. In: Proceedings of International Conference on Dependability of Computer Systems, DepCoS-RELCOMEX 2006, pp. 207 – 214 (2006). https://doi.org/10.1109/DEPCOS-RELCOMEX.2006.38
16. Alaee, S., Kamgar, K., Keogh, E.: Matrix profile XXII: Exact discovery of time series motifs under DTW. In: Proceedings - IEEE International Conference on Data Mining, ICDM, vol. 2020-November, pp. 900 – 905 (2020). https://doi.org/10.1109/ICDM50108.2020.00099
17. Alaee, S., Mercer, R., Kamgar, K., Keogh, E.: Time series motifs discovery under DTW allows more robust discovery of conserved structure. Data Mining and Knowledge Discovery **35**(3), 863 – 910 (2021). https://doi.org/10.1007/s10618-021-00740-0
18. Alamr, A., Artoli, A.: Unsupervised Transformer-Based Anomaly Detection in ECG Signals. Algorithms **16**(3) (2023). https://doi.org/10.3390/a16030152
19. Alevizos, E., Skarlatidis, A., Artikis, A., Paliouras, G.: Probabilistic complex event recognition: A survey. ACM Computing Surveys **50**(5) (2017). https://doi.org/10.1145/3117809
20. Alimohammadi, H., Nancy Chen, S.: Performance evaluation of outlier detection techniques in production timeseries: A systematic review and meta-analysis. Expert Systems with Applications **191** (2022). https://doi.org/10.1016/j.eswa.2021.116371
21. Almeida, A., Brás, S., Sargento, S., Pinto, F.C.: Time series big data: a survey on data stream frameworks, analysis and algorithms. Journal of Big Data **10**(1) (2023). https://doi.org/10.1186/s40537-023-00760-1
22. Amador Coelho, R., Bambirra Torres, L.C., Leite de Castro, C.: Concept drift detection with quadtree-based spatial mapping of streaming data. Information Sciences **625**, 578 – 592 (2023). https://doi.org/10.1016/j.ins.2022.12.085
23. Aminikhanghahi, S., Cook, D.J.: A survey of methods for time series change point detection. Knowledge and Information Systems **51**(2), 339 – 367 (2017). https://doi.org/10.1007/s10115-016-0987-z
24. An, X., Jiang, D., Liu, C., Zhao, M.: Wind farm power prediction based on wavelet decomposition and chaotic time series. Expert Systems with Applications **38**(9), 11,280 – 11,285 (2011). https://doi.org/10.1016/j.eswa.2011.02.176
25. Andrews, D.W.K.: Tests for Parameter Instability and Structural Change With Unknown Change Point. Econometrica **61**(4), 821–856 (1993). https://doi.org/10.2307/2951764
26. Andrienko, G., Andrienko, N., Mladenov, M., Mock, M., Poelitz, C.: Extracting events from spatial time series. In: Proceedings of the International Conference on Information Visualisation, pp. 48 – 53 (2010). https://doi.org/10.1109/IV.2010.17
27. Ané, T., Ureche-Rangau, L., Gambet, J.B., Bouverot, J.: Robust outlier detection for Asia-Pacific stock index returns. Journal of International Financial Markets, Institutions and Money **18**(4), 326 – 343 (2008). https://doi.org/10.1016/j.intfin.2007.03.001
28. Ang, Y., Huang, Q., Tung, A.K.H., Huang, Z.: A Stitch in Time Saves Nine: Enabling Early Anomaly Detection with Correlation Analysis. In: Proceedings - International Conference on Data Engineering, vol. 2023-April, pp. 1832 – 1845 (2023). https://doi.org/10.1109/ICDE55515.2023.00143

29. Anguera, A., Lara, J.A., Lizcano, D., Martínez, M.A., Pazos, J.: Sensor-generated time series events: A definition language. Sensors (Switzerland) **12**(9), 11,811 – 11,852 (2012). https://doi.org/10.3390/s120911811
30. Ansari, A.F., Stella, L., Turkmen, C., Zhang, X., Mercado, P., Shen, H., Shchur, O., Rangapuram, S.S., Arango, S.P., Kapoor, S., Zschiegner, J., Maddix, D.C., Wang, H., Mahoney, M.W., Torkkola, K., Wilson, A.G., Bohlke-Schneider, M., Wang, Y.: Chronos: Learning the Language of Time Series (2024). https://doi.org/10.48550/arXiv.2403.07815. http://arxiv.org/abs/2403.07815
31. Ares, J., Lara, J., Lizcano, D., Suarez, S.: A soft computing framework for classifying time series based on fuzzy sets of events. Information Sciences **330**, 125–144 (2016). https://doi.org/10.1016/j.ins.2015.10.014
32. Ariyaluran Habeeb, R.A., Nasaruddin, F., Gani, A., Targio Hashem, I.A., Ahmed, E., Imran, M.: Real-time big data processing for anomaly detection: A Survey. International Journal of Information Management **45**, 289 – 307 (2019). https://doi.org/10.1016/j.ijinfomgt.2018.08.006
33. Babüroğlu, E.S., Durmuşoğlu, A., Dereli, T.: Concept drift from 1980 to 2020: a comprehensive bibliometric analysis with future research insight. Evolving Systems (2023). https://doi.org/10.1007/s12530-023-09503-2
34. Barbará, D., Domeniconi, C., Rogers, J.P.: Detecting outliers using transduction and statistical testing. In: Proceedings of the 12th ACM SIGKDD international conference on Knowledge discovery and data mining, KDD '06, pp. 55–64. Association for Computing Machinery, New York, NY, USA (2006). https://doi.org/10.1145/1150402.1150413
35. Bayram, F., Ahmed, B.S., Kassler, A.: From concept drift to model degradation: An overview on performance-aware drift detectors. Knowledge-Based Systems **245** (2022). https://doi.org/10.1016/j.knosys.2022.108632
36. Benkabou, S.E., Benabdeslem, K., Canitia, B.: Unsupervised outlier detection for time series by entropy and dynamic time warping. Knowledge and Information Systems **54**(2), 463 – 486 (2018). https://doi.org/10.1007/s10115-017-1067-8
37. Bentkus, V.: On Hoeffding's inequalities. Annals of Probability **32**(2), 1650 – 1673 (2004). https://doi.org/10.1214/009117904000000360
38. Bifet, A., Gavaldà, R.: Learning from time-changing data with adaptive windowing. In: Proceedings of the 7th SIAM International Conference on Data Mining, pp. 443 – 448. Society for Industrial and Applied Mathematics Publications (2007). https://doi.org/10.1137/1.9781611972771.42
39. Blázquez-García, A., Conde, A., Mori, U., Lozano, J.A.: A Review on Outlier/Anomaly Detection in Time Series Data. ACM Computing Surveys **54**(3) (2021). https://doi.org/10.1145/3444690
40. Boniol, P., Palpanas, T.: Series2graph: Graph-based subsequence anomaly detection for time series. Proceedings of the VLDB Endowment **13**(11), 1821 – 1834 (2020). https://doi.org/10.14778/3407790.3407792
41. Boniol, P., Paparrizos, J., Kang, Y., Palpanas, T., Tsay, R.S., Elmore, A.J., Franklin, M.J.: Theseus: Navigating the Labyrinth of Time-Series Anomaly Detection. Proceedings of the VLDB Endowment **15**(12), 3702 – 3705 (2022). https://doi.org/10.14778/3554821.3554879
42. Boniol, P., Paparrizos, J., Palpanas, T.: New Trends in Time-Series Anomaly Detection. In: Advances in Database Technology - EDBT, vol. 26, pp. 847 – 850 (2023). https://doi.org/10.48786/edbt.2023.80
43. Boniol, P., Paparrizos, J., Palpanas, T., Franklin, M.J.: SAND: streaming subsequence anomaly detection. Proceedings of the VLDB Endowment **14**(10), 1717–1729 (2021). https://doi.org/10.14778/3467861.3467863

44. Borges, H., Dutra, M., Bazaz, A., Coutinho, R., Perosi, F., Porto, F., Masseglia, F., Pacitti, E., Ogasawara, E.: Spatial-time motifs discovery. Intelligent Data Analysis **24**(5), 1121 – 1140 (2020). https://doi.org/10.3233/IDA-194759
45. Box, G.E.P., Jenkins, G.M., Reinsel, G.C., Ljung, G.M.: Time Series Analysis: Forecasting and Control. John Wiley & Sons (2015)
46. Bozdogan, H.: Model selection and Akaike's Information Criterion (AIC): The general theory and its analytical extensions. Psychometrika **52**(3), 345 – 370 (1987). https://doi.org/10.1007/BF02294361
47. Bracewell, R.N., Bracewell, R.: The Fourier Transform and Its Applications. McGraw Hill (2000)
48. Braei, M., Wagner, S.: Anomaly Detection in Univariate Time-series: A Survey on the State-of-the-Art (2020). https://doi.org/10.48550/arXiv.2004.00433. http://arxiv.org/abs/2004.00433
49. Breiman, L.: Random forests. Machine Learning **45**(1), 5 – 32 (2001). https://doi.org/10.1023/A:1010933404324
50. Breuniq, M.M., Kriegel, H.P., Ng, R.T., Sander, J.: LOF: Identifying density-based local outliers. SIGMOD Record (ACM Special Interest Group on Management of Data) **29**(2), 93 – 104 (2000). https://doi.org/10.1145/335191.335388
51. Brockwell, P.J., Davis, R.A.: Introduction to Time Series and Forecasting. Springer International Publishing (2016)
52. Buda, T.S., Assem, H., Xu, L.: ADE: An ensemble approach for early Anomaly Detection. In: Proceedings of the IM 2017 - 2017 IFIP/IEEE International Symposium on Integrated Network and Service Management, pp. 442 – 448 (2017). https://doi.org/10.23919/INM.2017.7987310
53. Buhler, J., Tompa, M.: Finding motifs using random projections. Journal of Computational Biology **9**(2), 225 – 242 (2002). https://doi.org/10.1089/10665270252935430
54. Buza, K.: Time series classification and its applications. In: ACM International Conference Proceeding Series (2018). https://doi.org/10.1145/3227609.3227690
55. Carmona, R.: Statistical Analysis of Financial Data in R. Springer Science & Business Media (2013)
56. Carreño, A., Inza, I., Lozano, J.A.: Analyzing rare event, anomaly, novelty and outlier detection terms under the supervised classification framework. Artificial Intelligence Review **53**(5), 3575 – 3594 (2020). https://doi.org/10.1007/s10462-019-09771-y
57. Castro, N., Azevedo, P.: Multiresolution motif discovery in time series. In: Proceedings of the 10th SIAM International Conference on Data Mining, SDM 2010, pp. 665 – 676 (2010). https://doi.org/10.1137/1.9781611972801.73
58. Castro, N.C., Azevedo, P.J.: Significant motifs in time series. Statistical Analysis and Data Mining **5**(1), 35 – 53 (2012). https://doi.org/10.1002/sam.11134
59. Cerdà-Alabern, L., Iuhasz, G., Gemmi, G.: Anomaly detection for fault detection in wireless community networks using machine learning. Computer Communications **202**, 191 – 203 (2023). https://doi.org/10.1016/j.comcom.2023.02.019
60. Chandola, V., Banerjee, A., Kumar, V.: Anomaly detection: A survey. ACM Computing Surveys **41**(3) (2009). https://doi.org/10.1145/1541880.1541882
61. Chandola, V., Banerjee, A., Kumar, V.: Anomaly detection for discrete sequences: A survey. IEEE Transactions on Knowledge and Data Engineering **24**(5), 823 – 839 (2012). https://doi.org/10.1109/TKDE.2010.235
62. Chandola, V., Mithal, V., Kumar, V.: Comparative evaluation of anomaly detection techniques for sequence data. In: Proceedings - IEEE International Conference on Data Mining, ICDM, pp. 743 – 748 (2008). https://doi.org/10.1109/ICDM.2008.151

63. Chatzikonstanti, V.: Breaks and outliers when modelling the volatility of the U.S. stock market. Applied Economics **49**(46), 4704 – 4717 (2017). https://doi.org/10.1080/00036846.2017.1293785
64. Chawla, S., Gionisy, A.: k-means-: A unified approach to clustering and outlier detection. In: Proceedings of the 2013 SIAM International Conference on Data Mining, SDM 2013, pp. 189 – 197 (2013). https://doi.org/10.1137/1.9781611972832.21
65. Chen, H., Rossi, R.A., Mahadik, K., Kim, S., Eldardiry, H.: Graph Deep Factors for Probabilistic Time-series Forecasting. ACM Transactions on Knowledge Discovery from Data **17**(2) (2023). https://doi.org/10.1145/3543511
66. Chen, J., Gupta, A.K.: Parametric Statistical Change Point Analysis. Springer Science & Business Media (2013)
67. Chen, Q., Zhang, A., Huang, T., He, Q., Song, Y.: Imbalanced dataset-based echo state networks for anomaly detection. Neural Computing and Applications **32**(8), 3685–3694 (2020). https://doi.org/10.1007/s00521-018-3747-z
68. Chen, X., Huang, K., Jiang, H.: Detecting Changes in the Spatiotemporal Pattern of Bike Sharing: A Change-Point Topic Model. IEEE Transactions on Intelligent Transportation Systems **23**(10), 18,361 – 18,377 (2022). https://doi.org/10.1109/TITS.2022.3161623
69. Cheng, C., Sa-Ngasoongsong, A., Beyca, O., Le, T., Yang, H., Kong, Z., Bukkapatnam, S.T.: Time series forecasting for nonlinear and non-stationary processes: A review and comparative study. IIE Transactions (Institute of Industrial Engineers) **47**(10), 1053 – 1071 (2015). https://doi.org/10.1080/0740817X.2014.999180
70. Chicco, D., Warrens, M.J., Jurman, G.: The coefficient of determination R-squared is more informative than SMAPE, MAE, MAPE, MSE and RMSE in regression analysis evaluation. PeerJ Computer Science **7**, 1 – 24 (2021). https://doi.org/10.7717/PEERJ-CS.623
71. Chiu, B., Keogh, E., Lonardi, S.: Probabilistic discovery of time series motifs. In: Proceedings of the ACM SIGKDD International Conference on Knowledge Discovery and Data Mining, pp. 493 – 498 (2003). https://doi.org/10.1145/956750.956808
72. Chow, G.C.: Tests of Equality Between Sets of Coefficients in Two Linear Regressions. Econometrica **28**(3), 591–605 (1960). https://doi.org/10.2307/1910133
73. Clements, M.P., Hendry, D.F.: A Companion to Economic Forecasting. John Wiley & Sons (2008)
74. Coles, S.: An Introduction to Statistical Modeling of Extreme Values. Springer Science & Business Media (2013)
75. Conejo, A.J., Plazas, M.A., Espínola, R., Molina, A.B.: Day-ahead electricity price forecasting using the wavelet transform and ARIMA models. IEEE Transactions on Power Systems **20**(2), 1035 – 1042 (2005). https://doi.org/10.1109/TPWRS.2005.846054
76. Cook, A.A., Misirli, G., Fan, Z.: Anomaly Detection for IoT Time-Series Data: A Survey. IEEE Internet of Things Journal **7**(7), 6481 – 6494 (2020). https://doi.org/10.1109/JIOT.2019.2958185
77. Cormen, T.H., Leiserson, C.E., Rivest, R.L., Stein, C.: Introduction to Algorithms. MIT Press (2009)
78. Cryer, J.D., Chan, K.S.: Time Series Analysis: With Applications in R. Springer Science & Business Media (2008)
79. Dai, E., Chen, J.: Graph-Augmented Normalizing Flows for Anomaly Detection of Multiple Time Series (2022). https://doi.org/10.48550/arXiv.2202.07857. http://arxiv.org/abs/2202.07857
80. Dai, S., Meng, F.: Addressing modern and practical challenges in machine learning: a survey of online federated and transfer learning. Applied Intelligence **53**(9), 11,045–11,072 (2023). https://doi.org/10.1007/s10489-022-04065-3

81. Darban, Z.Z., Webb, G.I., Pan, S., Aggarwal, C.C., Salehi, M.: Deep Learning for Time Series Anomaly Detection: A Survey (2022). https://doi.org/10.48550/arXiv.2211.05244. http://arxiv.org/abs/2211.05244
82. Daw, C., Finney, C., Tracy, E.: A review of symbolic analysis of experimental data. Review of Scientific Instruments **74**(2), 915 – 930 (2003). https://doi.org/10.1063/1.1531823
83. De Paepe, D., Vanden Hautte, S., Steenwinckel, B., De Turck, F., Ongenae, F., Janssens, O., Van Hoecke, S.: A generalized matrix profile framework with support for contextual series analysis. Engineering Applications of Artificial Intelligence **90** (2020). https://doi.org/10.1016/j.engappai.2020.103487
84. De Ryck, T., De Vos, M., Bertrand, A.: Change Point Detection in Time Series Data Using Autoencoders with a Time-Invariant Representation. IEEE Transactions on Signal Processing **69**, 3513 – 3524 (2021). https://doi.org/10.1109/TSP.2021.3087031
85. Deng, A., Hooi, B.: Graph Neural Network-Based Anomaly Detection in Multivariate Time Series. In: 35th AAAI Conference on Artificial Intelligence, AAAI 2021, vol. 5A, pp. 4027 – 4035 (2021)
86. Diab, D.M., Assadhan, B., Binsalleeh, H., Lambotharan, S., Kyriakopoulos, K.G., Ghafir, I.: Anomaly detection using dynamic time warping. In: Proceedings - 22nd IEEE International Conference on Computational Science and Engineering and 17th IEEE International Conference on Embedded and Ubiquitous Computing, CSE/EUC 2019, pp. 193 – 198 (2019). https://doi.org/10.1109/CSE/EUC.2019.00045
87. Ding, H., Trajcevski, G., Scheuermann, P., Wang, X., Keogh, E.: Querying and mining of time series data: Experimental comparison of representations and distance measures. In: Proceedings of the VLDB Endowment, vol. 1, pp. 1542 – 1552 (2008). https://doi.org/10.14778/1454159.1454226
88. Ding, J., Xiang, Y., Shen, L., Tarokh, V.: Multiple Change Point Analysis: Fast Implementation and Strong Consistency. IEEE Transactions on Signal Processing **65**(17), 4495 – 4510 (2017). https://doi.org/10.1109/TSP.2017.2711558
89. Ditzler, G., Roveri, M., Alippi, C., Polikar, R.: Learning in Nonstationary Environments: A Survey. IEEE Computational Intelligence Magazine **10**(4), 12 – 25 (2015). https://doi.org/10.1109/MCI.2015.2471196
90. Dixit, P., Bhattacharya, P., Tanwar, S., Gupta, R.: Anomaly detection in autonomous electric vehicles using AI techniques: A comprehensive survey. Expert Systems **39**(5) (2022). https://doi.org/10.1111/exsy.12754
91. Dudek, G.: Neural networks for pattern-based short-term load forecasting: A comparative study. Neurocomputing **205**, 64 – 74 (2016). https://doi.org/10.1016/j.neucom.2016.04.021
92. Egilmez, H.E., Ortega, A.: Spectral anomaly detection using graph-based filtering for wireless sensor networks. In: ICASSP, IEEE International Conference on Acoustics, Speech and Signal Processing - Proceedings, pp. 1085 – 1089 (2014). https://doi.org/10.1109/ICASSP.2014.6853764
93. Elliott, G., Rothenberg, T.J., Stock, J.H.: Efficient tests for an autoregressive unit root. Econometrica **64**(4), 813 – 836 (1996). https://doi.org/10.2307/2171846
94. Ermshaus, A., Schäfer, P., Leser, U.: ClaSP: parameter-free time series segmentation. Data Mining and Knowledge Discovery **37**(3), 1262 – 1300 (2023). https://doi.org/10.1007/s10618-023-00923-x
95. Esling, P., Agon, C.: Time-series data mining. ACM Computing Surveys **45**(1) (2012). https://doi.org/10.1145/2379776.2379788
96. Ester, M., Kriegel, H.P., Sander, J., Xu, X.: A density-based algorithm for discovering clusters in large spatial databases with noise. In: Proceedings of the Second International Conference on

Knowledge Discovery and Data Mining, KDD'96, pp. 226–231. AAAI Press, Portland, Oregon (1996)
97. Fahrmann, D., Martin, L., Sanchez, L., Damer, N.: Anomaly Detection in Smart Environments: A Comprehensive Survey. IEEE Access **12**, 64,006 – 64,049 (2024). https://doi.org/10.1109/ACCESS.2024.3395051
98. Fanaee-T, H., Gama, J.: Event labeling combining ensemble detectors and background knowledge. Progress in Artificial Intelligence **2**(2-3), 113 – 127 (2014). https://doi.org/10.1007/s13748-013-0040-3
99. Flach, M., Gans, F., Brenning, A., Denzler, J., Reichstein, M., Rodner, E., Bathiany, S., Bodesheim, P., Guanche, Y., Sippel, S., Mahecha, M.D.: Multivariate anomaly detection for Earth observations: A comparison of algorithms and feature extraction techniques. Earth System Dynamics **8**(3), 677 – 696 (2017). https://doi.org/10.5194/esd-8-677-2017
100. Flandrin, P., Gonçalvès, P., Rilling, G.: Detrending and denoising with empirical mode decompositions. In: 2004 12th European Signal Processing Conference, pp. 1581–1584 (2004a)
101. Flandrin, P., Rilling, G., Gonçalvés, P.: Empirical mode decomposition as a filter bank. IEEE Signal Processing Letters **11**(2 PART I), 112 – 114 (2004b). https://doi.org/10.1109/LSP.2003.821662
102. Flórez, A., Rodríguez-Moreno, I., Artetxe, A., Olaizola, I.G., Sierra, B.: CatSight, a direct path to proper multi-variate time series change detection: perceiving a concept drift through common spatial pattern. International Journal of Machine Learning and Cybernetics **14**(9), 2925 – 2944 (2023). https://doi.org/10.1007/s13042-023-01810-z
103. Frías-Blanco, I., Del Campo-Ávila, J., Ramos-Jiménez, G., Morales-Bueno, R., Ortiz-Díaz, A., Caballero-Mota, Y.: Online and non-parametric drift detection methods based on Hoeffding's bounds. IEEE Transactions on Knowledge and Data Engineering **27**(3), 810 – 823 (2015). https://doi.org/10.1109/TKDE.2014.2345382
104. Fuchs, E., Gruber, T., Nitschke, J., Sick, B.: On-line motif detection in time series with Swift-Motif. Pattern Recognition **42**(11), 3015 – 3031 (2009). https://doi.org/10.1016/j.patcog.2009.05.004
105. Gabarda, S., Cristóbal, G.: Detection of events in seismic time series by time-frequency methods. IET Signal Processing **4**(4), 413 – 420 (2010). https://doi.org/10.1049/iet-spr.2009.0125
106. Gama, J., Medas, P., Castillo, G., Rodrigues, P.: Learning with drift detection. Lecture Notes in Computer Science (including subseries Lecture Notes in Artificial Intelligence and Lecture Notes in Bioinformatics) **3171**, 286 – 295 (2004). https://doi.org/10.1007/978-3-540-28645-5_29
107. Gama, J., Zliobaite, I., Bifet, A., Pechenizkiy, M., Bouchachia, A.: A survey on concept drift adaptation. ACM Computing Surveys **46**(4) (2014)https://doi.org/10.1145/2523813
108. Garcia, S., Luengo, J., Herrera, F.: Data Preprocessing in Data Mining. Springer (2014)
109. Geler, Z., Kurbalija, V., Ivanović, M., Radovanović, M.: Weighted kNN and constrained elastic distances for time-series classification. Expert Systems with Applications **162** (2020). https://doi.org/10.1016/j.eswa.2020.113829
110. Giusti, L., Carvalho, L., Gomes, A.T., Coutinho, R., Soares, J., Ogasawara, E.: Analyzing flight delay prediction under concept drift. Evolving Systems (2022). https://doi.org/10.1007/s12530-021-09415-z
111. Gonçalves, P.M., De Carvalho Santos, S.G., Barros, R.S., Vieira, D.C.: A comparative study on concept drift detectors. Expert Systems with Applications **41**(18), 8144 – 8156 (2014). https://doi.org/10.1016/j.eswa.2014.07.019
112. Górecki, T., Horváth, L., Kokoszka, P.: Change point detection in heteroscedastic time series. Econometrics and Statistics **7**, 63 – 88 (2018). https://doi.org/10.1016/j.ecosta.2017.07.005

113. Greff, K., Srivastava, R.K., Koutnik, J., Steunebrink, B.R., Schmidhuber, J.: LSTM: A Search Space Odyssey. IEEE Transactions on Neural Networks and Learning Systems **28**(10), 2222 – 2232 (2017). https://doi.org/10.1109/TNNLS.2016.2582924
114. Grubbs, F.E.: Procedures for Detecting Outlying Observations in Samples. Technometrics **11**(1), 1 – 21 (1969). https://doi.org/10.1080/00401706.1969.10490657
115. Gruver, N., Finzi, M., Qiu, S., Wilson, A.G.: Large Language Models Are Zero-Shot Time Series Forecasters. In: Advances in Neural Information Processing Systems, vol. 36 (2023)
116. Guerra-Manzanares, A., Bahsi, H.: On the relativity of time: Implications and challenges of data drift on long-term effective android malware detection. Computers and Security **122** (2022). https://doi.org/10.1016/j.cose.2022.102835
117. Guha, S., Mishra, N., Roy, G., Schrijvers, O.: Robust random cut forest based anomaly detection on streams. In: 33rd International Conference on Machine Learning, ICML 2016, vol. 6, pp. 3987 – 3999 (2016)
118. Guha, S., Rastogi, R., Shim, K.: Rock: a robust clustering algorithm for categorical attributes. Information Systems **25**(5), 345 – 366 (2000). https://doi.org/10.1016/S0306-4379(00)00022-3
119. Gujarati, D.N.: Essentials of Econometrics. SAGE (2021)
120. Guo, T., Dong, J., Li, H., Gao, Y.: Simple convolutional neural network on image classification. In: 2017 IEEE 2nd International Conference on Big Data Analysis, ICBDA 2017, pp. 721 – 724 (2017). https://doi.org/10.1109/ICBDA.2017.8078730
121. Gupta, M., Gao, J., Aggarwal, C.C., Han, J.: Outlier Detection for Temporal Data: A Survey. IEEE Transactions on Knowledge and Data Engineering **26**(9), 2250 – 2267 (2014). https://doi.org/10.1109/TKDE.2013.184
122. Gupta, M., Wadhvani, R., Rasool, A.: Real-time Change-Point Detection: A deep neural network-based adaptive approach for detecting changes in multivariate time series data. Expert Systems with Applications **209** (2022). https://doi.org/10.1016/j.eswa.2022.118260
123. Guralnik, V., Srivastava, J.: Event Detection from Time Series Data. In: Proceedings of the Fifth ACM SIGKDD International Conference on Knowledge Discovery and Data Mining, KDD ’99, pp. 33–42. ACM, New York, NY, USA (1999). https://doi.org/10.1145/312129.312190
124. Hahsler, M., Bolaños, M.: Clustering Data Streams Based on Shared Density between Micro-Clusters. IEEE Transactions on Knowledge and Data Engineering **28**(6), 1449–1461 (2016). https://doi.org/10.1109/TKDE.2016.2522412
125. Han, J., Pei, J., Tong, H.: Data Mining: Concepts and Techniques. 4th edition edn. Morgan Kaufmann, Cambridge, MA (2022a)
126. Han, M., Chen, Z., Li, M., Wu, H., Zhang, X.: A survey of active and passive concept drift handling methods. Computational Intelligence **38**(4), 1492 – 1535 (2022b). https://doi.org/10.1111/coin.12520
127. Han, S., Woo, S.S.: Learning Sparse Latent Graph Representations for Anomaly Detection in Multivariate Time Series. In: Proceedings of the ACM SIGKDD International Conference on Knowledge Discovery and Data Mining, pp. 2977 – 2986 (2022). https://doi.org/10.1145/3534678.3539117
128. Han, Z., Zhao, J., Leung, H., Ma, K.F., Wang, W.: A Review of Deep Learning Models for Time Series Prediction. IEEE Sensors Journal **21**(6), 7833 – 7848 (2021). https://doi.org/10.1109/JSEN.2019.2923982
129. Hanssens, D.M., Parsons, L.J., Schultz, R.L.: Market Response Models: Econometric and Time Series Analysis. Springer Science & Business Media (2012)
130. Haykin, S.O.: Neural Networks and Learning Machines. Pearson Education (2011)

131. Heim, N., Avery, J.E.: Adaptive Anomaly Detection in Chaotic Time Series with a Spatially Aware Echo State Network (2019). https://doi.org/10.48550/arXiv.1909.01709. http://arxiv.org/abs/1909.01709
132. Herbert, T., Mangler, J., Rinderle-Ma, S.: Generating Reliable Process Event Streams and Time Series Data Based on Neural Networks. Lecture Notes in Business Information Processing **421**, 81 – 95 (2021). https://doi.org/10.1007/978-3-030-79186-5_6
133. Hinkley, D.V.: Inference about the change-point in a sequence of random variables. Biometrika **57**(1), 1 – 17 (1970)https://doi.org/10.1093/biomet/57.1.1
134. Hlávka, Z., Hušková, M., Meintanis, S.G.: Change-point methods for multivariate time-series: paired vectorial observations. Statistical Papers **61**(4), 1351 – 1383 (2020). https://doi.org/10.1007/s00362-020-01175-3
135. Hoens, T., Polikar, R., Chawla, N.V.: Learning from streaming data with concept drift and imbalance: An overview. Progress in Artificial Intelligence **1**(1), 89 – 101 (2012). https://doi.org/10.1007/s13748-011-0008-0
136. Holder, C., Middlehurst, M., Bagnall, A.: A review and evaluation of elastic distance functions for time series clustering. Knowledge and Information Systems **66**(2), 765 – 809 (2024). https://doi.org/10.1007/s10115-023-01952-0
137. Huang, G.B., Zhu, Q.Y., Siew, C.K.: Extreme learning machine: Theory and applications. Neurocomputing **70**(1-3), 489 – 501 (2006). https://doi.org/10.1016/j.neucom.2005.12.126
138. Hundman, K., Constantinou, V., Laporte, C., Colwell, I., Soderstrom, T.: Detecting Spacecraft Anomalies Using LSTMs and Nonparametric Dynamic Thresholding. In: Proceedings of the 24th ACM SIGKDD International Conference on Knowledge Discovery & Data Mining, KDD '18, pp. 387–395. Association for Computing Machinery, New York, NY, USA (2018). https://doi.org/10.1145/3219819.3219845
139. Hyndman, R.J., Athanasopoulos, G.: Forecasting: principles and practice. OTexts (2018)
140. Ibidunmoye, O., Hernández-Rodriguez, F., Elmroth, E.: Performance anomaly detection and bottleneck identification. ACM Computing Surveys **48**(1) (2015). https://doi.org/10.1145/2791120
141. Iglesias Vázquez, F., Hartl, A., Zseby, T., Zimek, A.: Anomaly detection in streaming data: A comparison and evaluation study. Expert Systems with Applications **233** (2023). https://doi.org/10.1016/j.eswa.2023.120994
142. Iwashita, A.S., Papa, J.P.: An Overview on Concept Drift Learning. IEEE Access **7**, 1532 – 1547 (2019). https://doi.org/10.1109/ACCESS.2018.2886026
143. Jacob, V., Song, F., Stiegler, A., Rad, B., Diao, Y., Tatbul, N.: Exathlon: A benchmark for explainable anomaly detection over time series. Proceedings of the VLDB Endowment **14**(11), 2613 – 2626 (2021). https://doi.org/10.14778/3476249.3476307
144. Jain, A.K.: Data clustering: 50 years beyond K-means. Pattern Recognition Letters **31**(8), 651 – 666 (2010). https://doi.org/10.1016/j.patrec.2009.09.011
145. James, G.M., Witten, D., Hastie, T., Tibshirani, R.: An Introduction to Statistical Learning: With Applications in R. Springer Nature (2021)
146. Jensen, S.K., Pedersen, T.B., Thomsen, C.: Time Series Management Systems: A Survey. IEEE Transactions on Knowledge and Data Engineering **29**(11), 2581 – 2600 (2017). https://doi.org/10.1109/TKDE.2017.2740932
147. Jin, M., Wang, S., Ma, L., Chu, Z., Zhang, J.Y., Shi, X., Chen, P.Y., Liang, Y., Li, Y.F., Pan, S., Wen, Q.: Time-LLM: Time Series Forecasting by Reprogramming Large Language Models (2024). https://doi.org/10.48550/arXiv.2310.01728. http://arxiv.org/abs/2310.01728
148. Jollife, I.T., Cadima, J.: Principal component analysis: A review and recent developments. Philosophical Transactions of the Royal Society A: Mathematical, Physical and Engineering Sciences **374**(2065) (2016). https://doi.org/10.1098/rsta.2015.0202

149. Joo, T.W., Kim, S.B.: Time series forecasting based on wavelet filtering. Expert Systems with Applications **42**(8), 3868 – 3874 (2015). https://doi.org/10.1016/j.eswa.2015.01.026
150. Jouini, J.: Bootstrap methods for single structural change tests: Power versus corrected size and empirical illustration. Statistical Papers **51**(1), 85 – 109 (2010). https://doi.org/10.1007/s00362-008-0123-6
151. Kattan, A., Fatima, S., Arif, M.: Time-series event-based prediction: An unsupervised learning framework based on genetic programming. Information Sciences **301**, 99 – 123 (2015). https://doi.org/10.1016/j.ins.2014.12.054
152. Kaufman, L., Rousseeuw, P.J.: Finding Groups in Data: An Introduction to Cluster Analysis. John Wiley & Sons (2009)
153. Keogh, E.: The UCR Matrix Profile Page. Tech. rep., https://www.cs.ucr.edu/~eamonn/MatrixProfile.html (2024)
154. Keogh, E., Kasetty, S.: On the Need for Time Series Data Mining Benchmarks: A Survey and Empirical Demonstration. Data Mining and Knowledge Discovery **7**(4), 349 – 371 (2003). https://doi.org/10.1023/A:1024988512476
155. Keogh, E., Lonardi, S., Ratanamahatana, C.A.: Towards parameter-free data mining. In: KDD-2004 - Proceedings of the Tenth ACM SIGKDD International Conference on Knowledge Discovery and Data Mining, pp. 206 – 215 (2004). https://doi.org/10.1145/1014052.1014077
156. Keogh, E., Ratanamahatana, C.A.: Exact indexing of dynamic time warping. Knowledge and Information Systems **7**(3), 358 – 386 (2005). https://doi.org/10.1007/s10115-004-0154-9
157. Khamassi, I., Sayed-Mouchaweh, M., Hammami, M., Ghédira, K.: Discussion and review on evolving data streams and concept drift adapting. Evolving Systems **9**(1), 1 – 23 (2018). https://doi.org/10.1007/s12530-016-9168-2
158. Kieu, T., Yang, B., Jensen, C.S.: Outlier detection for multidimensional time series using deep neural networks. In: Proceedings - IEEE International Conference on Mobile Data Management, vol. 2018-June, pp. 125 – 134 (2018). https://doi.org/10.1109/MDM.2018.00029
159. Killick, R., Eckley, I.A.: Changepoint: An R package for changepoint analysis. Journal of Statistical Software **58**(3), 1 – 19 (2014). https://doi.org/10.18637/jss.v058.i03
160. Killick, R., Fearnhead, P., Eckley, I.: Optimal detection of changepoints with a linear computational cost. Journal of the American Statistical Association **107**(500), 1590 – 1598 (2012). https://doi.org/10.1080/01621459.2012.737745
161. Kim, C., Lee, J., Kim, R., Park, Y., Kang, J.: DeepNAP: Deep neural anomaly pre-detection in a semiconductor fab. Information Sciences **457-458**, 1–11 (2018). https://doi.org/10.1016/j.ins.2018.05.020
162. Kim, H., Kim, B., Chung, D., Yoon, J., Ko, S.K.: SoccerCPD: Formation and Role Change-Point Detection in Soccer Matches Using Spatiotemporal Tracking Data. In: Proceedings of the ACM SIGKDD International Conference on Knowledge Discovery and Data Mining, pp. 3146 – 3156 (2022). https://doi.org/10.1145/3534678.3539150
163. Kloska, M., Grmanova, G., Rozinajova, V.: Expert enhanced dynamic time warping based anomaly detection. Expert Systems with Applications **225** (2023). https://doi.org/10.1016/j.eswa.2023.120030
164. Lahmiri, S.: Interest rate next-day variation prediction based on hybrid feedforward neural network, particle swarm optimization, and multiresolution techniques. Physica A: Statistical Mechanics and its Applications **444**, 388 – 396 (2016). https://doi.org/10.1016/j.physa.2015.09.061
165. Lavin, A., Ahmad, S.: Evaluating real-time anomaly detection algorithms - The numenta anomaly benchmark. In: Proceedings - 2015 IEEE 14th International Conference on Machine Learning and Applications, ICMLA 2015, pp. 38 – 44 (2016). https://doi.org/10.1109/ICMLA.2015.141

166. Laxhammar, R., Falkman, G.: Inductive conformal anomaly detection for sequential detection of anomalous sub-trajectories. Annals of Mathematics and Artificial Intelligence **74**(1-2), 67 – 94 (2015). https://doi.org/10.1007/s10472-013-9381-7
167. Lazarevic, A., Kumar, V.: Feature bagging for outlier detection. In: Proceedings of the ACM SIGKDD International Conference on Knowledge Discovery and Data Mining, pp. 157 – 166 (2005). https://doi.org/10.1145/1081870.1081891
168. Lemire, D.: Faster retrieval with a two-pass dynamic-time-warping lower bound. Pattern Recognition **42**(9), 2169 – 2180 (2009). https://doi.org/10.1016/j.patcog.2008.11.030
169. Li, B., Müller, E.: STAD: State-Transition-Aware Anomaly Detection Under Concept Drifts. In: CEUR Workshop Proceedings, vol. 3380 (2022)
170. Li, J., Izakian, H., Pedrycz, W., Jamal, I.: Clustering-based anomaly detection in multivariate time series data. Applied Soft Computing **100** (2021). https://doi.org/10.1016/j.asoc.2020.106919
171. Li, J., Malialis, K., Polycarpou, M.M.: Autoencoder-based Anomaly Detection in Streaming Data with Incremental Learning and Concept Drift Adaptation. In: Proceedings of the International Joint Conference on Neural Networks, vol. 2023-June (2023). https://doi.org/10.1109/IJCNN54540.2023.10191328
172. Lim, B., Zohren, S.: Time-series forecasting with deep learning: A survey. Philosophical Transactions of the Royal Society A: Mathematical, Physical and Engineering Sciences **379**(2194) (2021). https://doi.org/10.1098/rsta.2020.0209
173. Lima, J., Salles, R., Porto, F., Coutinho, R., Alpis, P., Escobar, L., Pacitti, E., Ogasawara, E.: Forward and Backward Inertial Anomaly Detector: A Novel Time Series Event Detection Method. In: Proceedings of the International Joint Conference on Neural Networks, vol. 2022-July, pp. 1–8 (2022). https://doi.org/10.1109/IJCNN55064.2022.9892088
174. Lima, J., Tavares, L.G., Pacitti, E., Ferreira, J.E., Santos, I., Siqueira, I.G., Carvalho, D., Porto, F., Coutinho, R., Ogasawara, E.: Online Event Detection in Streaming Time Series: Novel Metrics and Practical Insights. In: Proceedings of the International Joint Conference on Neural Networks, vol. 2024-July, pp. 1–8 (2024)
175. Lin, J., Keogh, E., Lonardi, S., Chiu, B.: A symbolic representation of time series, with implications for streaming algorithms. In: Proceedings of the 8th ACM SIGMOD Workshop on Research Issues in Data Mining and Knowledge Discovery, DMKD '03, pp. 2 – 11 (2003). https://doi.org/10.1145/882082.882086
176. Lin, J., Keogh, E., Wei, L., Lonardi, S.: Experiencing SAX: A novel symbolic representation of time series. Data Mining and Knowledge Discovery **15**(2), 107 – 144 (2007). https://doi.org/10.1007/s10618-007-0064-z
177. Linardi, M., Zhu, Y., Palpanas, T., Keogh, E.: Matrix profile goes MAD: variable-length motif and discord discovery in data series. Data Mining and Knowledge Discovery **34**(4), 1022 – 1071 (2020). https://doi.org/10.1007/s10618-020-00685-w
178. Liu, A., Song, Y., Zhang, G., Lu, J.: Regional concept drift detection and density synchronized drift adaptation. In: IJCAI International Joint Conference on Artificial Intelligence, vol. 0, pp. 2280 – 2286 (2017). https://doi.org/10.24963/ijcai.2017/317
179. Liu, B., Zhao, H., Liu, Y., Wang, S., Li, J., Li, Y., Lang, J., Gu, R.: Discovering multi-dimensional motifs from multi-dimensional time series for air pollution control. Concurrency and Computation: Practice and Experience **32**(11) (2020). https://doi.org/10.1002/cpe.5645
180. Liu, F., Deng, Y.: A Fast Algorithm for Network Forecasting Time Series. IEEE Access **7**, 102,554 – 102,560 (2019). https://doi.org/10.1109/ACCESS.2019.2926986
181. Los, C.: Financial Market Risk: Measurement and Analysis. Routledge (2003)

182. Loureiro, D., Amado, C., Martins, A., Vitorino, D., Mamade, A., Coelho, S.T.: Water distribution systems flow monitoring and anomalous event detection: A practical approach. Urban Water Journal **13**(3), 242 – 252 (2016). https://doi.org/10.1080/1573062X.2014.988733
183. Lu, J., Liu, A., Dong, F., Gu, F., Gama, J., Zhang, G.: Learning under Concept Drift: A Review. IEEE Transactions on Knowledge and Data Engineering **31**(12), 2346 – 2363 (2019). https://doi.org/10.1109/TKDE.2018.2876857
184. Lughofer, E., Angelov, P.: Handling drifts and shifts in on-line data streams with evolving fuzzy systems. Applied Soft Computing Journal **11**(2), 2057 – 2068 (2011). https://doi.org/10.1016/j.asoc.2010.07.003
185. Ma, J., Perkins, S.: Time-series Novelty Detection Using One-class Support Vector Machines. In: Proceedings of the International Joint Conference on Neural Networks, vol. 3, pp. 1741 – 1745 (2003)
186. Machado, E., Serqueira, M., Ogasawara, E., Ogando, R., Maia, M.A.G., Da Costa, L.N., Campisano, R., Paiva Guedes, G., Bezerra, E.: Exploring machine learning methods for the Star/-Galaxy Separation Problem. In: Proceedings of the International Joint Conference on Neural Networks, vol. 2016-October, pp. 123 – 130 (2016). https://doi.org/10.1109/IJCNN.2016.7727189
187. Malhotra, P., Ramakrishnan, A., Anand, G., Vig, L., Agarwal, P., Shroff, G.: LSTM-based Encoder-Decoder for Multi-sensor Anomaly Detection (2016). https://doi.org/10.48550/arXiv.1607.00148. http://arxiv.org/abs/1607.00148
188. Manzoor, E., Lamba, H., Akoglu, L.: XStream: Outlier dete'X'ion in feature-evolving data streams. In: Proceedings of the ACM SIGKDD International Conference on Knowledge Discovery and Data Mining, pp. 1963 – 1972 (2018). https://doi.org/10.1145/3219819.3220107
189. Mao, S., Xiao, F.: Time Series Forecasting Based on Complex Network Analysis. IEEE Access **7**, 40,220 – 40,229 (2019). https://doi.org/10.1109/ACCESS.2019.2906268
190. McGovern, A., Rosendahl, D.H., Brown, R.A., Droegemeier, K.K.: Identifying predictive multi-dimensional time series motifs: An application to severe weather prediction. Data Mining and Knowledge Discovery **22**(1-2), 232 – 258 (2011). https://doi.org/10.1007/s10618-010-0193-7
191. Mehrmolaei, S., Keyvanpour, M.R.: A brief survey on event prediction methods in time series. Advances in Intelligent Systems and Computing **347**, 235 – 246 (2015). https://doi.org/10.1007/978-3-319-18476-0_24
192. Memarzadeh, M., Matthews, B., Avrekh, I.: Unsupervised anomaly detection in flight data using convolutional variational auto-encoder. Aerospace **7**(8) (2020). https://doi.org/10.3390/AEROSPACE7080115
193. Minnen, D., Isbell, C.L., Essa, I., Starner, T.: Discovering multivariate motifs using subsequence density estimation and greedy mixture learning. In: Proceedings of the National Conference on Artificial Intelligence, vol. 1, pp. 615 – 620 (2007)
194. Moghram, I., Rahman, S.: Analysis and evaluation of five short-term load forecasting techniques. IEEE Transactions on Power Systems **4**(4), 1484 – 1491 (1989). https://doi.org/10.1109/59.41700
195. Moghtaderi, A., Borgnat, P., Flandrin, P.: Trend filtering: Empirical mode decompositions versus l1 and hodrick-prescott. Advances in Adaptive Data Analysis **3**(1-2), 41 – 61 (2011). https://doi.org/10.1142/S1793536911000751
196. Molaei, S.M., Keyvanpour, M.R.: An analytical review for event prediction system on time series. In: 2015 2nd International Conference on Pattern Recognition and Image Analysis, IPRIA 2015, pp. 1–6 (2015). https://doi.org/10.1109/PRIA.2015.7161635
197. Morana, C.: Multivariate modelling of long memory processes with common components. Computational Statistics and Data Analysis **52**(2), 919 – 934 (2007). https://doi.org/10.1016/j.csda.2006.12.010

198. Mueen, A.: Enumeration of time series motifs of all lengths. In: Proceedings - IEEE International Conference on Data Mining, ICDM, pp. 547 – 556 (2013). https://doi.org/10.1109/ICDM.2013.27
199. Mueen, A.: Time series motif discovery: Dimensions and applications. Wiley Interdisciplinary Reviews: Data Mining and Knowledge Discovery **4**(2), 152 – 159 (2014). https://doi.org/10.1002/widm.1119
200. Mueen, A., Keogh, E., Zhu, Q., Cash, S., Westover, B.: Exact discovery of time series motifs. In: Society for Industrial and Applied Mathematics - 9th SIAM International Conference on Data Mining 2009, Proceedings in Applied Mathematics, vol. 1, pp. 469 – 480 (2009)
201. Mueen, A., Keogh, E., Zhu, Q., Cash, S.S., Westover, M.B., Bigdely-Shamlo, N.: A disk-aware algorithm for time series motif discovery. Data Mining and Knowledge Discovery **22**(1-2), 73 – 105 (2011). https://doi.org/10.1007/s10618-010-0176-8
202. Munir, M., Chattha, M.A., Dengel, A., Ahmed, S.: A comparative analysis of traditional and deep learning-based anomaly detection methods for streaming data. In: Proceedings - 18th IEEE International Conference on Machine Learning and Applications, ICMLA 2019, pp. 561 – 566 (2019a). https://doi.org/10.1109/ICMLA.2019.00105
203. Munir, M., Siddiqui, S.A., Dengel, A., Ahmed, S.: DeepAnT: A Deep Learning Approach for Unsupervised Anomaly Detection in Time Series. IEEE Access **7**, 1991 – 2005 (2019b). https://doi.org/10.1109/ACCESS.2018.2886457
204. NASA: Global Warming vs. Climate Change. Tech. rep., https://climate.nasa.gov/global-warming-vs-climate-change/ (2023)
205. Nelson, C.R., Plosser, C.R.: Trends and random walks in macroeconmic time series. Some evidence and implications. Journal of Monetary Economics **10**(2), 139 – 162 (1982). https://doi.org/10.1016/0304-3932(82)90012-5
206. NOAA: Climate at a Glance Global Time Series. Tech. rep., https://www.ncei.noaa.gov/access/monitoring/climate-at-a-glance/global/time-series (2023)
207. Ntroumpogiannis, A., Giannoulis, M., Myrtakis, N., Christophides, V., Simon, E., Tsamardinos, I.: A meta-level analysis of online anomaly detectors. VLDB Journal **32**(4), 845 – 886 (2023). https://doi.org/10.1007/s00778-022-00773-x
208. Nunthanid, P., Niennattrakul, V., Ratanamahatana, C.A.: Discovery of variable length time series motif. In: ECTI-CON 2011 - 8th Electrical Engineering/ Electronics, Computer, Telecommunications and Information Technology (ECTI) Association of Thailand - Conference 2011, pp. 472 – 475 (2011). https://doi.org/10.1109/ECTICON.2011.5947877
209. Ogasawara, E., Martinez, L.C., De Oliveira, D., Zimbrão, G., Pappa, G.L., Mattoso, M.: Adaptive Normalization: A novel data normalization approach for non-stationary time series. In: Proceedings of the International Joint Conference on Neural Networks (2010). https://doi.org/10.1109/IJCNN.2010.5596746
210. Ogasawara, E., Murta, L., Zimbrão, G., Mattoso, M.: Neural networks cartridges for data mining on time series. In: Proceedings of the International Joint Conference on Neural Networks, pp. 2302 – 2309 (2009). https://doi.org/10.1109/IJCNN.2009.5178615
211. Ogasawara, E., Salles, R., Lima, J., Baroni, L., Castro, A., Carvalho, L., Borges, H., Carvalho, D., Coutinho, R., Bezerra, E., Pacitti, E., Porto, F.: harbinger: A Unified Time Series Event Detection Framework (2023). https://cran.r-project.org/web/packages/harbinger/index.html
212. Olteanu, M., Rossi, F., Yger, F.: Meta-survey on outlier and anomaly detection. Neurocomputing **555** (2023). https://doi.org/10.1016/j.neucom.2023.126634
213. Ozkan, H., Ozkan, F., Kozat, S.S.: Online Anomaly Detection Under Markov Statistics With Controllable Type-I Error. IEEE Transactions on Signal Processing **64**(6), 1435 – 1445 (2016). https://doi.org/10.1109/TSP.2015.2504345

214. Pang, G., Shen, C., Cao, L., Van Den Hengel, A.: Deep Learning for Anomaly Detection: A Review. ACM Computing Surveys **54**(2) (2021). https://doi.org/10.1145/3439950
215. Paparrizos, J., Boniol, P., Palpanas, T., Tsay, R.S., Elmore, A., Franklin, M.J.: Volume Under the Surface: A New Accuracy Evaluation Measure for Time-Series Anomaly Detection. Proceedings of the VLDB Endowment **15**(11), 2774 – 2787 (2022a). https://doi.org/10.14778/3551793.3551830
216. Paparrizos, J., Kang, Y., Boniol, P., Tsay, R.S., Palpanas, T., Franklin, M.J.: TSB-UAD: An End-to-End Benchmark Suite for Univariate Time-Series Anomaly Detection. Proceedings of the VLDB Endowment **15**(8), 1697 – 1711 (2022b). https://doi.org/10.14778/3529337.3529354
217. Paparrizos, J., Liu, C., Elmore, A.J., Franklin, M.J.: Debunking Four Long-Standing Misconceptions of Time-Series Distance Measures. In: Proceedings of the 2020 ACM SIGMOD International Conference on Management of Data, SIGMOD '20, pp. 1887–1905. Association for Computing Machinery, New York, NY, USA (2020). https://doi.org/10.1145/3318464.3389760
218. Paparrizos, J., Wu, K., Elmore, A., Faloutsos, C., Franklin, M.J.: Accelerating Similarity Search for Elastic Measures: A Study and New Generalization of Lower Bounding Distances. Proceedings of the VLDB Endowment **16**(8), 2019–2032 (2023). https://doi.org/10.14778/3594512.3594530
219. Park, P., Di Marco, P., Shin, H., Bang, J.: Fault detection and diagnosis using combined autoencoder and long short-term memory network. Sensors (Switzerland) **19**(21) (2019). https://doi.org/10.3390/s19214612
220. Parra, L., Deco, G., Miesbach, S.: Statistical Independence and Novelty Detection with Information Preserving Nonlinear Maps. Neural Computation **8**(2), 260 – 269 (1996). https://doi.org/10.1162/neco.1996.8.2.260
221. Patel, P., Keogh, E., Lin, J., Lonardi, S.: Mining motifs in massive time series databases. In: Proceedings - IEEE International Conference on Data Mining, ICDM, pp. 370 – 377 (2002)
222. Percival, D.B., Walden, A.T.: Wavelet Methods for Time Series Analysis. Cambridge University Press (2006)
223. Pereira, R., Souto, Y., Chaves, A., Zorilla, R., Tsan, B., Rusu, F., Ogasawara, E., Ziviani, A., Porto, F.: DJEnsemble: A Cost-Based Selection and Allocation of a Disjoint Ensemble of Spatio-Temporal Models. In: ACM International Conference Proceeding Series, pp. 226 – 231 (2021). https://doi.org/10.1145/3468791.3468806
224. Pesaranghader, A., Viktor, H.L.: Fast hoeffding drift detection method for evolving data streams. Lecture Notes in Computer Science (including subseries Lecture Notes in Artificial Intelligence and Lecture Notes in Bioinformatics) **9852 LNAI**, 96 – 111 (2016). https://doi.org/10.1007/978-3-319-46227-1_7
225. Pesaranghader, A., Viktor, H.L., Paquet, E.: McDiarmid Drift Detection Methods for Evolving Data Streams. In: Proceedings of the International Joint Conference on Neural Networks, vol. 2018-July (2018). https://doi.org/10.1109/IJCNN.2018.8489260
226. Pevný, T.: Loda: Lightweight on-line detector of anomalies. Machine Learning **102**(2), 275 – 304 (2016). https://doi.org/10.1007/s10994-015-5521-0
227. PhysioNet: Records in the MIT-BIH Arrhythmia Database. Tech. rep., https://physionet.org/files/mitdb/1.0.0/mitdbdir/records.htm (2024)
228. Pimentel, M.A., Clifton, D.A., Clifton, L., Tarassenko, L.: A review of novelty detection. Signal Processing **99**, 215 – 249 (2014). https://doi.org/10.1016/j.sigpro.2013.12.026
229. Ploberger, W., Krämer, W.: The Cusum Test with Ols Residuals. Econometrica **60**(2), 271–285 (1992). https://doi.org/10.2307/2951597
230. Poiriernherbeck, L., Lahalle, E., Saurel, N., Marcos, S.: Unknown-length motif discovery methods in environmental monitoring time series. In: International Conference on Electrical, Com-

puter, and Energy Technologies, ICECET 2022 (2022). https://doi.org/10.1109/ICECET55527.2022.9873093
231. Porto, F., Ferro, M., Ogasawara, E., Moeda, T., de Barros, C.D.T., Silva, A.C., Zorrilla, R., Pereira, R.S., Castro, R.N., Silva, J.V., et al.: Machine learning approaches to extreme weather events forecast in urban areas: Challenges and initial results. Supercomputing Frontiers and Innovations **9**(1), 49–73 (2022)
232. Prabuchandran, K., Singh, N., Dayama, P., Agarwal, A., Pandit, V.: Change point detection for compositional multivariate data. Applied Intelligence **52**(2), 1930 – 1955 (2022). https://doi.org/10.1007/s10489-021-02321-6
233. Pratama, M., Lu, J., Lughofer, E., Zhang, G., Er, M.J.: An Incremental Learning of Concept Drifts Using Evolving Type-2 Recurrent Fuzzy Neural Networks. IEEE Transactions on Fuzzy Systems **25**(5), 1175 – 1192 (2017). https://doi.org/10.1109/TFUZZ.2016.2599855
234. Provotar, O.I., Linder, Y.M., Veres, M.M.: Unsupervised Anomaly Detection in Time Series Using LSTM-Based Autoencoders. In: 2019 IEEE International Conference on Advanced Trends in Information Theory, ATIT 2019 - Proceedings, pp. 513 – 517 (2019). https://doi.org/10.1109/ATIT49449.2019.9030505
235. Pyle, D.: Data Preparation for Data Mining. Morgan Kaufmann (1999)
236. Raab, C., Heusinger, M., Schleif, F.M.: Reactive Soft Prototype Computing for Concept Drift Streams. Neurocomputing **416**, 340 – 351 (2020). https://doi.org/10.1016/j.neucom.2019.11.111
237. Rakthanmanon, T., Campana, B., Mueen, A., Batista, G., Westover, B., Zhu, Q., Zakaria, J., Keogh, E.: Addressing big data time series: Mining trillions of time series subsequences under dynamic time warping. ACM Transactions on Knowledge Discovery from Data **7**(3) (2013). https://doi.org/10.1145/2500489
238. Rayana, S., Akoglu, L.: Less is more: Building selective anomaly ensembles. ACM Transactions on Knowledge Discovery from Data **10**(4) (2016). https://doi.org/10.1145/2890508
239. Ren, H., Xu, B., Wang, Y., Yi, C., Huang, C., Kou, X., Xing, T., Yang, M., Tong, J., Zhang, Q.: Time-series anomaly detection service at Microsoft. In: Proceedings of the ACM SIGKDD International Conference on Knowledge Discovery and Data Mining, pp. 3009 – 3017 (2019). https://doi.org/10.1145/3292500.3330680
240. Ren, J., Xia, F., Lee, I., Noori Hoshyar, A., Aggarwal, C.: Graph Learning for Anomaly Analytics: Algorithms, Applications, and Challenges. ACM Transactions on Intelligent Systems and Technology **14**(2) (2023). https://doi.org/10.1145/3570906
241. Rettig, L., Khayati, M., Cudre-Mauroux, P., Piorkowski, M.: Online anomaly detection over Big Data streams. In: Proceedings - 2015 IEEE International Conference on Big Data, IEEE Big Data 2015, pp. 1113 – 1122 (2015). https://doi.org/10.1109/BigData.2015.7363865
242. Ross, G.J., Adams, N.M., Tasoulis, D.K., Hand, D.J.: Exponentially weighted moving average charts for detecting concept drift. Pattern Recognition Letters **33**(2), 191 – 198 (2012). https://doi.org/10.1016/j.patrec.2011.08.019
243. Ruff, L., Kauffmann, J.R., Vandermeulen, R.A., Montavon, G., Samek, W., Kloft, M., Dietterich, T.G., Muller, K.R.: A Unifying Review of Deep and Shallow Anomaly Detection. Proceedings of the IEEE **109**(5), 756 – 795 (2021). https://doi.org/10.1109/JPROC.2021.3052449
244. Sakurai, G.Y., Lopes, J.F., Zarpelão, B.B., Barbon Junior, S.: Benchmarking Change Detector Algorithms from Different Concept Drift Perspectives. Future Internet **15**(5) (2023). https://doi.org/10.3390/fi15050169
245. Salfner, F., Lenk, M., Malek, M.: A survey of online failure prediction methods. ACM Computing Surveys **42**(3) (2010). https://doi.org/10.1145/1670679.1670680
246. Salles, R., Assis, L., Guedes, G., Bezerra, E., Porto, F., Ogasawara, E.: A framework for benchmarking machine learning methods using linear models for univariate time series prediction.

In: Proceedings of the International Joint Conference on Neural Networks, vol. 2017-May, pp. 2338 – 2345 (2017). https://doi.org/10.1109/IJCNN.2017.7966139
247. Salles, R., Belloze, K., Porto, F., Gonzalez, P.H., Ogasawara, E.: Nonstationary time series transformation methods: An experimental review. Knowledge-Based Systems **164**, 274 – 291 (2019). https://doi.org/10.1016/j.knosys.2018.10.041
248. Salles, R., Escobar, L., Baroni, L., Zorrilla, R., Ziviani, A., Kreischer, V., Delicato, F., Pires, P.F., Maia, L., Coutinho, R., Assis, L., Ogasawara, E.: Harbinger: Um framework para integração e análise de métodos de detecção de eventos em séries temporais. In: Anais do Simpósio Brasileiro de Banco de Dados (SBBD), pp. 73–84. SBC (2020). https://doi.org/10.5753/sbbd.2020.13626
249. Salles, R., Lima, J., Coutinho, R., Pacitti, E., Masseglia, F., Akbarinia, R., Chen, C., Garibaldi, J., Porto, F., Ogasawara, E.: SoftED: Metrics for Soft Evaluation of Time Series Event Detection, Computers & Industrial Engineering,198, pp. 1 – 14, (2024). https://doi.org/10.1016/j.cie.2024.110728
250. Salles, R., Pacitti, E., Bezerra, E., Porto, F., Ogasawara, E.: TSPred: A framework for nonstationary time series prediction. Neurocomputing **467**, 197 – 202 (2022). https://doi.org/10.1016/j.neucom.2021.09.067
251. Sayed-Mouchaweh, M., Lughofer, E.: Learning in Non-Stationary Environments: Methods and Applications. Springer Science & Business Media (2012)
252. Schäfer, P., Leser, U.: Motiflets - Simple and Accurate Detection of Motifs in Time Series. Proceedings of the VLDB Endowment **16**(4), 725–737 (2022). https://doi.org/10.14778/3574245.3574257
253. Schmidl, S., Wenig, P., Papenbrock, T.: Anomaly Detection in Time Series: A Comprehensive Evaluation. Proceedings of the VLDB Endowment **15**(9), 1779 – 1797 (2022). https://doi.org/10.14778/3538598.3538602
254. Sebestyen, G., Hangan, A., Czako, Z., Kovacs, G.: A taxonomy and platform for anomaly detection. In: 2018 IEEE International Conference on Automation, Quality and Testing, Robotics, AQTR 2018 - THETA 21st Edition, Proceedings, pp. 1 – 6 (2018). https://doi.org/10.1109/AQTR.2018.8402710
255. Seth, T., Chaudhary, V.: A Predictive Analytics Framework for Insider Trading Events. In: Proceedings - 2020 IEEE International Conference on Big Data, Big Data 2020, pp. 218 – 225 (2020). https://doi.org/10.1109/BigData50022.2020.9377791
256. Shalev-Shwartz, S., Ben-David, S.: Understanding Machine Learning: From Theory to Algorithms. Cambridge University Press, USA (2014)
257. Shi, X., Gallagher, C., Lund, R., Killick, R.: A comparison of single and multiple changepoint techniques for time series data. Computational Statistics and Data Analysis **170** (2022). https://doi.org/10.1016/j.csda.2022.107433
258. Shrimankar, D.D.: High performance computing approach for DNA motif discovery. CSI Transactions on ICT **7**(4), 295–297 (2019). https://doi.org/10.1007/s40012-019-00235-w
259. Shumway, R.H., Stoffer, D.S.: Time Series Analysis and Its Applications: With R Examples. Springer (2017)
260. Siffer, A., Fouque, P.A., Termier, A., Largouet, C.: Anomaly detection in streams with extreme value theory. In: Proceedings of the ACM SIGKDD International Conference on Knowledge Discovery and Data Mining, vol. Part F129685, pp. 1067 – 1075 (2017). https://doi.org/10.1145/3097983.3098144
261. Singh, N., Olinsky, C.: Demystifying Numenta anomaly benchmark. In: Proceedings of the International Joint Conference on Neural Networks, vol. 2017-May, pp. 1570 – 1577 (2017). https://doi.org/10.1109/IJCNN.2017.7966038

262. Son, N.T., Anh, D.T.: Discovery of time series k-motifs based on multidimensional index. Knowledge and Information Systems **46**(1), 59 – 86 (2016). https://doi.org/10.1007/s10115-014-0814-3
263. Song, H., Jiang, Z., Men, A., Yang, B.: A Hybrid Semi-Supervised Anomaly Detection Model for High-Dimensional Data. Computational Intelligence and Neuroscience **2017**(1), 8501,683 (2017). https://doi.org/10.1155/2017/8501683
264. Sorbo, S., Ruocco, M.: Navigating the metric maze: a taxonomy of evaluation metrics for anomaly detection in time series. Data Mining and Knowledge Discovery **38**(3), 1027 – 1068 (2024). https://doi.org/10.1007/s10618-023-00988-8
265. Souto, G., Liebig, T.: On event detection from spatial time series for urban traffic applications. Lecture Notes in Computer Science (including subseries Lecture Notes in Artificial Intelligence and Lecture Notes in Bioinformatics) **9580**, 221 – 233 (2016). https://doi.org/10.1007/978-3-319-41706-6_11
266. Souza, J., Paixão, E., Fraga, F., Baroni, L., Alves, R.F.S., Belloze, K., Santos, J., Bezerra, E., Porto, F., Ogasawara, E.: REMD: A Novel Hybrid Anomaly Detection Method Based on EMD and ARIMA. In: Proceedings of the International Joint Conference on Neural Networks, vol. 2024-July, pp. 1–8 (2024)
267. Stefanakos, C., Schinas, O.: Forecasting bunker prices; a nonstationary, multivariate methodology. Transportation Research Part C: Emerging Technologies **38**, 177 – 194 (2014). https://doi.org/10.1016/j.trc.2013.11.017
268. Su, Y., Liu, R., Zhao, Y., Sun, W., Niu, C., Pei, D.: Robust anomaly detection for multivariate time series through stochastic recurrent neural network. In: Proceedings of the ACM SIGKDD International Conference on Knowledge Discovery and Data Mining, pp. 2828 – 2837 (2019). https://doi.org/10.1145/3292500.3330672
269. Sun, G., Chen, T., Wei, Z., Sun, Y., Zang, H., Chen, S.: A carbon price forecasting model based on variational mode decomposition and spiking neural networks. Energies **9**(1) (2016). https://doi.org/10.3390/en9010054
270. Taha, A., Hadi, A.S.: Anomaly detection methods for categorical data: A review. ACM Computing Surveys **52**(2) (2019). https://doi.org/10.1145/3312739
271. Takahashi, C.C., Braga, A.P.: A Review of Off-Line Mode Dataset Shifts. IEEE Computational Intelligence Magazine **15**(3), 16 – 27 (2020). https://doi.org/10.1109/MCI.2020.2998231
272. Takeuchi, J.I., Yamanishi, K.: A unifying framework for detecting outliers and change points from time series. IEEE Transactions on Knowledge and Data Engineering **18**(4), 482 – 492 (2006). https://doi.org/10.1109/TKDE.2006.1599387
273. Talagala, P.D., Hyndman, R.J., Smith-Miles, K., Kandanaarachchi, S., Muñoz, M.A.: Anomaly Detection in Streaming Nonstationary Temporal Data. Journal of Computational and Graphical Statistics **29**(1), 13 – 27 (2020). https://doi.org/10.1080/10618600.2019.1617160
274. Tanaka, Y., Iwamoto, K., Uehara, K.: Discovery of time-series motif from multi-dimensional data based on MDL principle. Machine Learning **58**(2-3), 269 – 300 (2005). https://doi.org/10.1007/s10994-005-5829-2
275. Tang, H., Liao, S.S.: Discovering original motifs with different lengths from time series. Knowledge-Based Systems **21**(7), 666 – 671 (2008). https://doi.org/10.1016/j.knosys.2008.03.022
276. Tatbul, N., Lee, T.J., Zdonik, S., Alam, M., Gottschlich, J.: Precision and recall for time series. In: Advances in Neural Information Processing Systems, vol. 2018-December, pp. 1920 – 1930 (2018)
277. Thudumu, S., Branch, P., Jin, J., Singh, J.: A comprehensive survey of anomaly detection techniques for high dimensional big data. Journal of Big Data **7**(1) (2020). https://doi.org/10.1186/s40537-020-00320-x

278. Tiao, G.: Asymptotic behaviour of temporal aggregates of time series. Biometrika **59**(3), 525 – 531 (1972). https://doi.org/10.1093/biomet/59.3.525
279. Torkamani, S., Lohweg, V.: Survey on time series motif discovery. Wiley Interdisciplinary Reviews: Data Mining and Knowledge Discovery **7**(2) (2017). https://doi.org/10.1002/widm.1199
280. Tran, L., Mun, M.Y., Shahabi, C.: Real-time distance-based outlier detection in data streams. Proceedings of the VLDB Endowment **14**(2), 141 – 153 (2020). https://doi.org/10.14778/3425879.3425885
281. Truong, C., Oudre, L., Vayatis, N.: Selective review of offline change point detection methods. Signal Processing **167** (2020). https://doi.org/10.1016/j.sigpro.2019.107299
282. Truong, C.D., Anh, D.T.: An efficient method for motif and anomaly detection in time series based on clustering. International Journal of Business Intelligence and Data Mining **10**(4), 356 – 377 (2015). https://doi.org/10.1504/IJBIDM.2015.072212
283. Tsay, R.S.: Analysis of Financial Time Series. Wiley (2010)
284. Vahdatpour, A., Amini, N., Sarrafzadeh, M.: Toward unsupervised activity discovery using multi-dimensional motif detection in time series. In: IJCAI International Joint Conference on Artificial Intelligence, pp. 1261 – 1266 (2009)
285. Van Onsem, M., Ledoux, V., Melange, W., Dreesen, D., Van Hoecke, S.: Variable Length Motif Discovery in Time Series Data. IEEE Access **11**, 73,754 – 73,766 (2023). https://doi.org/10.1109/ACCESS.2023.3295995
286. Vargas, R.E.V., Munaro, C.J., Ciarelli, P.M. , Medeiros, A.G. , do Amaral, B.G., Barrionuevo, D.C., de Araújo, J.C.D. , Ribeiro, J.L., and Magalhães, L.P.: A Realistic and Public Dataset with Rare Undesirable Real Events in Oil Wells, Journal of Petroleum Science and Engineering, **181**, pp. 1– 9 (2019). https://doi.org/10.1016/j.petrol.2019.106223
287. Vishwakarma, G.K., Paul, C., Elsawah, A.: An algorithm for outlier detection in a time series model using backpropagation neural network. Journal of King Saud University - Science **32**(8), 3328 – 3336 (2020). https://doi.org/10.1016/j.jksus.2020.09.018
288. Wang, C., Viswanathan, K., Choudur, L., Talwar, V., Satterfield, W., Schwan, K.: Statistical techniques for online anomaly detection in data centers. In: Proceedings of the 12th IFIP/IEEE International Symposium on Integrated Network Management, IM 2011, pp. 385 – 392 (2011). https://doi.org/10.1109/INM.2011.5990537
289. Wang, H., Bah, M.J., Hammad, M.: Progress in Outlier Detection Techniques: A Survey. IEEE Access **7**, 107,964 – 108,000 (2019). https://doi.org/10.1109/ACCESS.2019.2932769
290. Wang, H., Liu, L., Dong, S., Qian, Z., Wei, H.: A novel work zone short-term vehicle-type specific traffic speed prediction model through the hybrid EMD-ARIMA framework. Transportmetrica B **4**(3), 159 – 186 (2016). https://doi.org/10.1080/21680566.2015.1060582
291. Wang, L., Chng, E., Li, H.: A tree-construction search approach for multivariate time series motifs discovery. Pattern Recognition Letters **31**(9), 869 – 875 (2010). https://doi.org/10.1016/j.patrec.2010.01.005
292. Wang, P., Jin, N., Davies, D., Woo, W.L.: Model-centric transfer learning framework for concept drift detection. Knowledge-Based Systems **275** (2023). https://doi.org/10.1016/j.knosys.2023.110705
293. Ward, J.A., Lukowicz, P., Gellersen, H.W.: Performance metrics for activity recognition. ACM Transactions on Intelligent Systems and Technology **2**(1) (2011). https://doi.org/10.1145/1889681.1889687
294. Webb, G.I., Hyde, R., Cao, H., Nguyen, H.L., Petitjean, F.: Characterizing concept drift. Data Mining and Knowledge Discovery **30**(4), 964 – 994 (2016). https://doi.org/10.1007/s10618-015-0448-4

295. Wei, W.W.S.: Time Series Analysis Univariate and Multivariate Methods. Pearson Education (2018)
296. Wenig, P., Schmidl, S., Papenbrock, T.: TimeEval: a benchmarking toolkit for time series anomaly detection algorithms. Proceedings of the VLDB Endowment **15**(12), 3678–3681 (2022). https://doi.org/10.14778/3554821.3554873
297. Wenig, P., Schmidl, S., Papenbrock, T.: Anomaly Detectors for Multivariate Time Series: The Proof of the Pudding is in the Eating. In: Proceedings - 2024 IEEE 40th International Conference on Data Engineering Workshops, ICDEW 2024, pp. 96 – 101 (2024). https://doi.org/10.1109/ICDEW61823.2024.00018
298. Witten, I.H., Frank, E., Hall, M.A., Pal, C.J.: Data Mining: Practical Machine Learning Tools and Techniques. Morgan Kaufmann (2016)
299. Wu, R., Keogh, E.J.: Current Time Series Anomaly Detection Benchmarks are Flawed and are Creating the Illusion of Progress. IEEE Transactions on Knowledge and Data Engineering **35**(3), 2421 – 2429 (2023). https://doi.org/10.1109/TKDE.2021.3112126
300. Wu, W., He, L., Lin, W., Su, Y., Cui, Y., Maple, C., Jarvis, S.: Developing an Unsupervised Real-Time Anomaly Detection Scheme for Time Series With Multi-Seasonality. IEEE Transactions on Knowledge and Data Engineering **34**(9), 4147–4160 (2022). https://doi.org/10.1109/TKDE.2020.3035685
301. Wu, Z., Huang, N.E., Long, S.R., Peng, C.K.: On the trend, detrending, and variability of nonlinear and nonstationary time series. Proceedings of the National Academy of Sciences of the United States of America **104**(38), 14,889 – 14,894 (2007). https://doi.org/10.1073/pnas.0701020104
302. Xiang, Q., Zi, L., Cong, X., Wang, Y.: Concept Drift Adaptation Methods under the Deep Learning Framework: A Literature Review. Applied Sciences (Switzerland) **13**(11) (2023). https://doi.org/10.3390/app13116515
303. Xiao, Q., Si, Y.: Time series prediction using graph model. In: 2017 3rd IEEE International Conference on Computer and Communications, ICCC 2017, vol. 2018-January, pp. 1358 – 1361 (2018). https://doi.org/10.1109/CompComm.2017.8322764
304. Xie, F., Song, A., Ciesielski, V.: Event detection in time series by genetic programming. In: 2012 IEEE Congress on Evolutionary Computation, CEC 2012 (2012). https://doi.org/10.1109/CEC.2012.6256589
305. Xuan, P.T., Anh, D.T.: An efficient hash-based method for time series motif discovery. In: Multi-disciplinary Trends in Artificial Intelligence, vol. 11248 LNAI, pp. 205 – 211 (2018). https://doi.org/10.1007/978-3-030-03014-8_17
306. Yadav, R., Pradhan, A.K., Kamwa, I.: Real-Time Multiple Event Detection and Classification in Power System Using Signal Energy Transformations. IEEE Transactions on Industrial Informatics **15**(3), 1521 – 1531 (2019). https://doi.org/10.1109/TII.2018.2855428
307. Yang, J., Wang, W., Yu, P.S.: Mining surprising periodic patterns. Data Mining and Knowledge Discovery **9**(2), 189 – 216 (2004). https://doi.org/10.1023/B:DAMI.0000031631.84034.af
308. Yang, W., Zurbenko, I.: Nonstationarity. Wiley Interdisciplinary Reviews: Computational Statistics **2**(1), 107 – 115 (2010). https://doi.org/10.1002/wics.64
309. Yao, Y., Sharma, A., Golubchik, L., Govindan, R.: Online anomaly detection for sensor systems: A simple and efficient approach. Performance Evaluation **67**(11), 1059–1075 (2010). https://doi.org/10.1016/j.peva.2010.08.018
310. Yeh, C.C.M., Kavantzas, N., Keogh, E.: Matrix profile VI: Meaningful multidimensional motif discovery. In: Proceedings - IEEE International Conference on Data Mining, ICDM, vol. 2017-November, pp. 565 – 574 (2017a). https://doi.org/10.1109/ICDM.2017.66
311. Yeh, C.C.M., Zhu, Y., Ulanova, L., Begum, N., Ding, Y., Dau, H.A., Silva, D.F., Mueen, A., Keogh, E.: Matrix profile I: All pairs similarity joins for time series: A unifying view that

includes motifs, discords and shapelets. In: Proceedings - IEEE International Conference on Data Mining, ICDM, pp. 1317 – 1322 (2017b). https://doi.org/10.1109/ICDM.2016.89
312. Yeh, C.C.M., Zhu, Y., Ulanova, L., Begum, N., Ding, Y., Dau, H.A., Zimmerman, Z., Silva, D.F., Mueen, A., Keogh, E.: Time series joins, motifs, discords and shapelets: a unifying view that exploits the matrix profile. Data Mining and Knowledge Discovery **32**(1), 83 – 123 (2018). https://doi.org/10.1007/s10618-017-0519-9
313. Yeo, I.K., Johnson, R.A.: A new family of power transformations to improve normality or symmetry. Biometrika **87**(4), 954 – 959 (2000). https://doi.org/10.1093/biomet/87.4.954
314. You, X., Zhang, M., Ding, D., Feng, F., Huang, Y.: Learning to Learn the Future: Modeling Concept Drifts in Time Series Prediction. In: International Conference on Information and Knowledge Management, Proceedings, pp. 2434 – 2443 (2021). https://doi.org/10.1145/3459637.3482271
315. Yu, Q., Luo, Y., Chen, C., Wang, X.: Trajectory outlier detection approach based on common slices sub-sequence. Applied Intelligence **48**(9), 2661 – 2680 (2018). https://doi.org/10.1007/s10489-017-1104-z
316. Yuan, L., Li, H., Xia, B., Gao, C., Liu, M., Yuan, W., You, X.: Recent Advances in Concept Drift Adaptation Methods for Deep Learning. In: IJCAI International Joint Conference on Artificial Intelligence, pp. 5654 – 5661 (2022)
317. Zeileis, A.: A unified approach to structural change tests based on ML scores, F statistics, and OLS residuals. Econometric Reviews **24**(4), 445 – 466 (2005). https://doi.org/10.1080/07474930500406053
318. Zeileis, A., Kleiber, C., Walter, K., Hornik, K.: Testing and dating of structural changes in practice. Computational Statistics and Data Analysis **44**(1-2), 109 – 123 (2003). https://doi.org/10.1016/S0167-9473(03)00030-6
319. Zeileis, A., Leisch, F., Hornik, K., Kleiber, C.: Strucchange: An R package for testing for structural change in linear regression models. Journal of Statistical Software **7**, 1 – 38 (2002). https://doi.org/10.18637/jss.v007.i02
320. Zhang, C., Song, D., Chen, Y., Feng, X., Lumezanu, C., Cheng, W., Ni, J., Zong, B., Chen, H., Chawla, N.V.: A deep neural network for unsupervised anomaly detection and diagnosis in multivariate time series data. In: 33rd AAAI Conference on Artificial Intelligence, AAAI 2019, pp. 1409 – 1416 (2019)
321. Zhang, J., Wu, C., Wang, Y.: Human fall detection based on body posture spatio-temporal evolution. Sensors (Switzerland) **20**(3) (2020). https://doi.org/10.3390/s20030946
322. Zhang, Y., Meratnia, N., Havinga, P.: Outlier detection techniques for wireless sensor networks: A survey. IEEE Communications Surveys and Tutorials **12**(2), 159 – 170 (2010). https://doi.org/10.1109/SURV.2010.021510.00088
323. Zhao, H., Wang, Y., Duan, J., Huang, C., Cao, D., Tong, Y., Xu, B., Bai, J., Tong, J., Zhang, Q.: Multivariate time-series anomaly detection via graph attention network. In: Proceedings - IEEE International Conference on Data Mining, ICDM, vol. 2020-November, pp. 841 – 850 (2020a). https://doi.org/10.1109/ICDM50108.2020.00093
324. Zhao, J., Mo, H., Deng, Y.: An Efficient Network Method for Time Series Forecasting Based on the DC Algorithm and Visibility Relation. IEEE Access **8**, 7598 – 7608 (2020b). https://doi.org/10.1109/ACCESS.2020.2964067
325. Zhao, L.: Event Prediction in the Big Data Era: A Systematic Survey. ACM Computing Surveys **54**(5) (2021). https://doi.org/10.1145/3450287
326. Zhao, Q., Zhang, Y., Shi, Y., Li, J.: Analyzing and Visualizing Anomalies and Events in Time Series of Network Traffic. Advances in Intelligent Systems and Computing **936**, 15 – 25 (2020c). https://doi.org/10.1007/978-3-030-19861-9_2

327. Zhou, Q., Chen, J., Liu, H., He, S., Meng, W.: Detecting Multivariate Time Series Anomalies with Zero Known Label. In: Proceedings of the 37th AAAI Conference on Artificial Intelligence, AAAI 2023, vol. 37, pp. 4963 – 4971 (2023)
328. Zhou, Y., Arghandeh, R., Zou, H., Spanos, C.J.: Nonparametric Event Detection in Multiple Time Series for Power Distribution Networks. IEEE Transactions on Industrial Electronics **66**(2), 1619 – 1628 (2019). https://doi.org/10.1109/TIE.2018.2840508
329. Zhou, Y., Ren, H., Li, Z., Wu, N., Al-Ahmari, A.M.: Anomaly detection via a combination model in time series data. Applied Intelligence **51**(7), 4874 – 4887 (2021). https://doi.org/10.1007/s10489-020-02041-3
330. Zhu, Y., Yeh, C.C.M., Zimmerman, Z., Kamgar, K., Keogh, E.: Matrix Profile XI: SCRIMP++: Time Series Motif Discovery at Interactive Speeds. In: Proceedings - IEEE International Conference on Data Mining, ICDM, vol. 2018-November, pp. 837 – 846 (2018). https://doi.org/10.1109/ICDM.2018.00099
331. Zipfel, J., Verworner, F., Fischer, M., Wieland, U., Kraus, M., Zschech, P.: Anomaly detection for industrial quality assurance: A comparative evaluation of unsupervised deep learning models. Computers and Industrial Engineering **177** (2023). https://doi.org/10.1016/j.cie.2023.109045

Index

E. Ogasawara et al., *Event Detection in Time Series*, Synthesis Lectures on Data Management, https://doi.org/10.1007/978-3-031-75941-3